Contents

Introduction

Whatever type of job you do, or whatever technical subject you study, you will find that you need to be able to calculate quickly and accurately and to use certain mathematical techniques. This textbook is designed to help you by providing a comprehensive course in basic mathematics for a wide range of technical applications. It follows the order of the level-1 TEC mathematics syllabus (U75/005), but it is not necessary for you to work through the book in the order given.

The text is divided into 30 parts, each of which ends with a set of exercises. You are not expected to attempt all these questions, but it is advisable to try some which are relevant to the work you are doing. The questions are graded so that the later ones are harder than the ones at the start of an exercise.

At the end of the book is a set of revision questions which covers the work of the entire syllabus and illustrates the type of questions which might be set in a TEC examination in mathematics at level 1.

Follow-on books are being prepared for mathematics at level 2 and level 3, and there is also a companion book to this on physical science at level 1, by the same author.

A   Manipulating numbers

1 Integers, fractions and basic arithmetical operations

1.1 Integers, multiples and factors

An *integer* is a whole number.

The sequence 1, 2, 3, 4, . . . is the way we learn to count *positive* integers. The corresponding series of *negative* integers would be −1, −2, −3, −4, . . . but between these two sets of numbers we need to put in a value, neither positive nor negative, which we call *zero* or *nought* and for which we use the symbol 0.

If we add any two integers together, we get another integer:

e.g.   7 + 5 = 12

If we subtract one integer from another, we also get an integral number:

e.g.   7 − 5 = 2

If we add zero to an integer, or subtract zero from it, the integer remains unchanged:

e.g.   7 + 0 = 7

and   7 − 0 = 7

If we multiply any two integers together, we get another integer:

e.g.   7 x 5 = 35

Since we get 35 by multiplying 7 and 5 together, we say that 35 is the *product* of 7 and 5. Also, 35 is a *multiple* of 7 since 35 contains five sevens. Similarly, 35 must be a multiple of 5 since it contains seven fives. We call 7 and 5 *factors* of 35, which implies that they each divide into 35 without leaving any remainder.

Prime numbers

We can write 5 = 5 x 1 because multiplying anything by one (unity) leaves it unchanged. It follows that 5 and 1 are both factors of 5. Any integer which has no factors other than itself and unity is called a *prime* number. The only even prime number is 2, since every other even number is a multiple of 2. Thus 2, 5, 13, 47 and 101 are all examples of prime numbers.

A prime number which is a factor of another number is called a *prime factor*.

It is sometimes necessary to find out whether a particular integer is a prime number or, if it is not, to determine its factors. To do this it is usually best to take the sequence of prime numbers and divide each one in turn, as many times as possible, into the given integer.

Example   Express in prime factors the numbers 2310, 1989 and 2003.

2	2310		3	1989	2003
3	1155		3	663	is prime
5	385		13	221	
7	77		17	17	
11	11			1	
	1				

$\therefore$ $2310 = 2 \times 3 \times 5 \times 7 \times 11$ $1989 = 3^2 \times 13 \times 17$

Note in the above example that both 2310 and 1989 are divisible by 3. In such a case we say that 3 is a *common factor* of both.

HCF and LCM

The *highest common factor* (HCF) of two numbers is the largest number which will divide exactly into each of them.

The *lowest common multiple* (LCM) of two numbers is the smallest number of which both are factors.

Example   Find the HCF and LCM of 60 and 84.

$60 = 2 \times 2 \times 3 \times 5$
$84 = 2 \times 2 \times 3 \times 7$
HCF is $2 \times 2 \times 3 = 12$
LCM is $2 \times 2 \times 3 \times 5 \times 7 = 420$

Exercise A1

1. Copy out the following block of numbers on to a sheet of paper, then add the rows horizontally and the columns vertically. Check the totals by confirming that the resulting row and column of totals sum to the same grand total.

	Column 1	Column 2	Column 3	Column 4	Totals
Row (a)	347	743	473	437	
Row (b)	219	192	291	921	
Row (c)	726	627	267	762	
Row (d)	495	954	945	459	

Totals

2. Add the rows and columns and check the totals on the following block of numbers

23	3	11	−6	2
16	−2	24	10	0
9	17	0	21	−30
27	5	8	−7	25
4	13	31	0	18

3. Write down all the multiples of 7 between 25 and 50.
4. Which integers between 50 and 100 are divisible by 9?
5. Express in prime factors each of the following numbers:

   (a) 42   (b) 216   (c) 1024   (d) 1584

6. Write down the prime numbers between 20 and 30.
7. What is the next prime number after 47?
8. What factor do 15, 25 and 40 have in common?
9. Find the highest common factor of 96 and 144.
10. What is the lowest common multiple of 63 and 105?
11. Find the HCF and LCM of 135 and 225.
12. What is the HCF of 84, 112 and 168?
13. Explain why zero cannot be a factor.
14. Four lamp-posts are in a straight line along an esplanade at the sea-front. From the first lamp-post, the distances to the other three are 72 m, 108 m and 162 m respectively. Other lamp-posts are to be erected between the four existing ones to form a line of lights equally spaced. Find the appropriate spacing between adjacent lamp-posts if the smallest possible number of new ones is to be used.
15. One gear wheel with 48 teeth drives another with 54 teeth. Two teeth (one on each gear wheel) which start off next to one another are marked. Find how many times each gear wheel must turn before the marked teeth are back together in the same position.

1.2 Commutative laws

Numbers can be added together in any order without affecting the result, thus

$$6 + 7 = 13 \quad \text{and} \quad 7 + 6 = 13$$

It follows that

$$6 + 7 = 7 + 6$$

and in general

$$a + b = b + a$$

where the letter a stands for any number and b for another number. This is known as *the commutative law for addition*.

There is a corresponding rule called *the commutative law for multiplication* which can be written as

$$ab = ba$$

This implies that numbers can be multiplied together in any order without affecting the result, so that $6 \times 7 = 42$ and $7 \times 6 = 42$.

We conclude that addition and multiplication are both commutative.

It is important to note that many other operations are not commutative. We find that $8 \div 4$ and $4 \div 8$ do not give the same result, so division is not commutative. Similarly, $8 - 4$ and $4 - 8$ give different results, so subtraction is not commutative either.

3

1.3 Associative laws

These laws are concerned, not with the order in which numbers are placed, but with the order in which operations on the numbers are carried out. We find that addition is associative and a series of additions can be done in any order, e.g. $3 + (4 + 5) = (3 + 4) + 5$. In general, if we use letter a to represent the first number, b for the second, c for the third, then we can write

$$a + (b + c) = (a + b) + c$$

and this is known as *the associative law for addition.*

The corresponding *associative law for multiplication* can be written as

$$a(bc) = (ab)c$$

which implies that when two (or more) operations of multiplication are to be performed, these operations can be carried out in any order.

Many other operations are not associative, and it is important that the sequence in which the operations are to be carried out should be followed through exactly. (See also section 1.11 for precedence rules.)

Exercise   Check that $12 - (4 - 5)$ and $(12 - 4) - 5$ are not the same, and that $12 \div (4 \div 5)$ is different from $(12 \div 4) \div 5$.

Our results show that subtraction and division are not associative.

1.4 Distributive laws

We know that £1.05 = £1 + 5p. If the price of a certain article is £1.05 we can see that the total cost of three such articles would be £3.15, for $3 \times £1.05 = £3.15$. This implies that $3(£1 + 5p) = £3.15$, so what we actually do in finding $3 \times £1.05$ is to find $3 \times £1$ and $3 \times 5p$ and then add these products to get the total.

In general,

$$a(b + c) = ab + ac$$

and this is known as *the distributive law for multiplication.*

You may recognise it in this form as the rule for expanding brackets in algebra (which appears in section 10 later in this book) but distributive laws apply also to arithmetic and calculus. One example of the use of this distributive law would be in the quick calculation of 4×998. Since $998 = (1000 - 2)$, we see that

$$4 \times 998 = 4(1000 - 2) = 4000 - 8 = 3992$$

The distributive law for division may be stated as

$$\frac{a + b}{c} = \frac{a}{c} + \frac{b}{c} \quad \text{(where } c \text{ is not zero)}$$

This can be derived from the distributive law for multiplication by the same process, by which we can say that

$$\frac{a + b}{3} = \frac{1}{3}(a + b) = \frac{a}{3} + \frac{b}{3}$$

4

This law has obvious applications in adding and subtracting fractions (see section 1.8), but it is also used in algebra and calculus.

Warning   Some of the commonest errors in mathematics arise from attempts to apply a distributive law to an operation which is not distributive. You will avoid a lot of mistakes if you take note of the following examples.

a)   *Powers.*   It is true that $(a \times b)^2 = a^2 \times b^2$, but $(a + b)^2$ and $a^2 + b^2$ are not the same.

b)   *Roots.*   It is true that $\sqrt{a \times b} = \sqrt{a} \times \sqrt{b}$ but $\sqrt{a-b}$ and $\sqrt{a} - \sqrt{b}$ are not the same.

c)   *Trigonometrical functions* (see section G).   Sin $(A + B)$ is not the same as sin A + sin B and sin ax is not the same as sin $a \times$ sin x.

Check   If you are in any doubt about any operation or whether it is valid to apply a commutative, associative or distributive law, then it is always wise to check by substituting easy numbers into the relationship and seeing whether your operation gives the right answer or not.

Example   Are $(a - b)^2$ and $a^2 - b^2$ the same?
For a check, put $a = 3$, $b = 2$,

then $(a - b)^2 = (3 - 2)^2 = 1^2 = 1$

and $a^2 - b^2 = 3^2 - 2^2 = 9 - 4 = 5$

and it is clear that $(a - b)^2$ and $a^2 - b^2$ are not the same.

Exercise A2
1. For the law $a + b = b + a$, put $a = 5$, $b = 3$ and show that the law is true for this pair of numbers.
2. Use the values $a = 11$ and $b = 4$ to verify that $ab = ba$.
3. Verify the law $a + b + c = a + c + b$ when $a = 5$, $b = 3$ and $c = 8$.
4. Substitute the values $a = 10$, $b = 1$ and $c = 4$ to verify that $a + (b + c) = (a + b) + c$.
5. Use the values $a = 5$, $b = 3$ to show that $a - b$ and $b - a$ are not the same.
6. If $a = 2$, $b = 3$, $c = 5$, is $a(bc)$ the same as $(ab)c$?
7. Put in the values $a = 9$, $b = 4$, $c = 2$ to check whether $a + (b - c)$ is the same as $(a + b) - c$.
8. For the values $a = 5$, $b = 12$, $c = 4$, is $(a \times b) - c$ the same as $a \times (b - c)$?
9. By substituting $a = 2$, $b = 3$, $c = 4$, verify $a(b + c) = ab + ac$.
10. Use the law $a(b - c) = ab - ac$ to calculate the value of 7×999 by two different methods and show that the answer is the same in each case. (Hint: put $a = 7$, $b = 1000$, $c = 1$.)
11. Using the values $a = 5$, $b = 2$, $c = 6$, $d = 4$, see if $(a + b) + (c - d)$ is the same as $a + (b + c) - d$.
12. Check whether $\dfrac{a - b}{c + d}$ is the same as $\dfrac{a}{c} - \dfrac{b}{d}$ or $\dfrac{a}{c + d} - \dfrac{b}{c + d}$ for the values $a = 12$, $b = 8$, $c = 3$, $d = 1$.

1.5 Fractions

A fraction such as two-thirds may be written as $\frac{2}{3}$ or 2/3 or $2 \div 3$, but wherever we have more than one such ordinary or 'vulgar' fraction it is preferable to use the form $\frac{2}{3}$ to avoid any confusion when performing arithmetical operations involving fractions. In this form, with one number over another, we call the top number the *numerator* and the bottom number the *denominator*. We may wish to distinguish between *proper fractions,* in which the numerator is *less* than the denominator, and *improper fractions* which are 'top-heavy' in that the numerator is *greater* than the denominator. Thus $\frac{4}{5}$ and $\frac{13}{16}$ are proper fractions, but $\frac{5}{4}$ and $\frac{8}{3}$ are improper fractions. Any improper fraction can be reduced to a *mixed number* (an integer plus a proper fraction) by dividing the denominator into the numerator. Thus $\frac{5}{4} = 1\frac{1}{4}$ and $\frac{8}{3} = 2\frac{2}{3}$.

1.6 Cancellation

What fraction of an hour is 15 minutes? In realising that 15 minutes is equivalent to a quarter of an hour, we are equating $\frac{15}{60}$ with $\frac{1}{4}$. Similarly, $\frac{45}{60} = \frac{3}{4}$ because 45 minutes is three-quarters of an hour. We can reduce $\frac{45}{60}$ to $\frac{3}{4}$ all at once if we realise that 15 is the HCF of 45 and 60. Usually we would divide smaller factors progressively into both the numerator and the denominator of any fraction we wished to reduce, and if we divided first by 3 and then by 5 we would get

$$\frac{45}{60} = \frac{15}{20} = \frac{3}{4}$$

This process of reducing a fraction to its simplest form is known as *cancellation.*

If you spent the 24 hours of a particular day as follows: 8 sleeping, 9 working, 2 eating, 3 watching TV, 2 on other activities, you could put these in fractional form as $\frac{8}{24}$, $\frac{9}{24}$, $\frac{2}{24}$, $\frac{3}{24}$, $\frac{2}{24}$ and these would cancel down to $\frac{1}{3}$, $\frac{3}{8}$, $\frac{1}{12}$, $\frac{1}{8}$ and $\frac{1}{12}$ respectively.

1.7 Zero

Any fraction represents a ratio between the two parts which are the numerator and the denominator of the fraction. This ratio remains unaltered if both the numerator and the denominator are divided (or multiplied) by any factor other than zero. To cancel $\frac{9}{24}$ to $\frac{3}{8}$, we can visualize

$$\frac{9}{24} = \frac{3 \times 3}{8 \times 3}$$

so that it becomes obvious that 3 is a common factor which will cancel to leave

$$\frac{9}{24} = \frac{3}{8}$$

If we write a for *any* factor, $3a/8a$ becomes 3/8, since

$$\frac{3a}{8a} = \frac{3 \times a}{8 \times a} = \frac{3}{8}$$

6

The one exception to this, which we must guard against, is the possibility of *a* being zero. It is an extremely important rule in mathematics that we must *never cancel or divide by zero.*

1.8 Manipulation of fractions

For *addition* and *subtraction* it is necessary to use a *common denominator.* This common denominator should be the LCM of the denominators of the fractions being combined.

Example

$$\frac{1}{2} + \frac{2}{3} - \frac{3}{4} = \frac{6 + 8 - 9}{12} = \frac{5}{12}$$

When adding or subtracting mixed numbers, it is not necessary to turn them into improper fractions since it is easier to part the integers from the fractions and deal with them separately.

Example

$$4\tfrac{1}{3} - 2\tfrac{2}{5} + 3\tfrac{4}{15} = (4 - 2 + 3) + (\tfrac{1}{3} - \tfrac{2}{5} + \tfrac{4}{15})$$

$$= 5 + \frac{5 - 6 + 4}{15}$$

$$= 5 + \tfrac{3}{15}$$

$$= 5\tfrac{1}{5}$$

For *multiplication* of fractions we combine two operations, which can be done in any order. One operation is multiplication, and our final fraction is given by the product of all the numerators over the product of all the denominators. The other operation is cancellation, and we can cancel a factor of any numerator if we can find the same factor to cancel in any denominator.

Example $\dfrac{4}{9} \times \dfrac{5}{8} \times \dfrac{3}{5}$

If we multiply out first and then cancel we get

$$\frac{4}{9} \times \frac{5}{8} \times \frac{3}{5} = \frac{4 \times 5 \times 3}{9 \times 8 \times 5} = \frac{60}{360} = \frac{1}{6}$$

If we cancel first (by 3, 4 and 5) and then multiply we get

$$\frac{4}{9} \times \frac{5}{8} \times \frac{3}{5} = \frac{1}{3} \times \frac{1}{2} \times \frac{1}{1} = \frac{1}{6}$$

When multiplying mixed numbers it is necessary to change any mixed number to an improper fraction first (except for the simple case of a mixed number being multiplied by an integer).

7

Example

$$1\tfrac{9}{16} \times 2\tfrac{2}{15} = \frac{25}{16} \times \frac{32}{15} \quad \text{(cancel by 5 and 16)}$$

$$= \frac{5}{1} \times \frac{2}{3} = \frac{10}{3} = 3\tfrac{1}{3}$$

It should be noted that the word 'of' in this context is mathematically equivalent to a multiplication sign.

Example Calculate $\tfrac{2}{3}$ of $3\tfrac{3}{4}$

This is equivalent to $\tfrac{2}{3} \times \tfrac{15}{4}$ (cancel by 2 and 3):

$$\frac{1}{1} \times \frac{5}{2} = 2\tfrac{1}{2}$$

For *division* of fractions the standard procedure is to invert the divisor and multiply. Cancellation is used wherever possible and it is again necessary to change any mixed number to an improper fraction.

Example $\quad 2\tfrac{6}{7} \div 1\tfrac{4}{21} = \frac{20}{7} \div \frac{25}{21}$

$$= \frac{20}{7} \times \frac{21}{25} \quad \text{(cancel by 5 and 7)}$$

$$= \frac{4}{1} \times \frac{3}{5} = \frac{12}{5} = 2\tfrac{2}{5}$$

1.9 Ratio and proportion

Consider a map drawn to a scale such that 20 mm on the map represents 1 km on the ground. Any dimension, such as the length of a section of road, is reduced in the same proportion and, since 1 km = 1000 m = 1 000 000 mm, we find that lengths on the map are proportional to lengths on the ground in the *proportion* 20 : 1 000 000, i.e. 1 : 50 000. The *ratio* of map length to actual length is thus 1/50 000. This scale value is sometimes expressed in fractional form, so in this case we could say that the *representative fraction* (RF) of the map would be $\frac{1}{50\,000}$.

The close connection between the ideas of ratio, proportion and fraction is similarly involved in metallurgy when we consider alloys. If a certain alloy contains copper, manganese and nickel in the proportions 20:3:1 by weight, then 20 + 3 + 1 gives us 24 parts of which 20 will be copper, 3 will be manganese and 1 will be nickel. The fractional constituents are thus $\frac{20}{24} = \frac{5}{6}$ copper, $\frac{3}{24} = \frac{1}{8}$ manganese, and $\frac{1}{24}$ nickel. Three kilograms of finished alloy will therefore contain $\frac{5}{6} \times 3000$ g = 2500 g of copper, $\frac{1}{8} \times 3000$ g = 375 g of manganese and $\frac{1}{24} \times 3000$ g = 125 g of nickel. (Check: 2500 + 375 + 125 = 3000 g.)

It is important, both in mathematics and in science, to remember that a ratio has no dimensions. To find a ratio between two quantities it is essential to ensure that both quantities are in the same units. Conversely, the same ratio can be used to find the proportion between units even if they are different from those used to calculate the ratio originally. An aircraft may be built from plans which give dimensions in metres, but a scale model of the aircraft would be built to the same proportions but using different dimensions. The full-size plane and its small-scale model would have the same ratio of body length to wing span. Similarly, the ratio of US dollars to £ sterling is the same as the ratio of cents to pence.

1.10 Problems involving fractions

Example 1 Addition and subtraction
A tailor has a 20 m length of cloth. For making up various garments he cuts off the following lengths: $2\frac{1}{2}$ m, $3\frac{1}{4}$ m, $3\frac{3}{4}$ m and $1\frac{3}{4}$ m. How much is left?
The remaining length will be

$$20 - 2\tfrac{1}{2} - 3\tfrac{1}{4} - 3\tfrac{3}{4} - 1\tfrac{3}{4} = 11 - \frac{2+1+3+3}{4}$$

$$= 11 - \tfrac{9}{4} = 11 - 2\tfrac{1}{4}$$

$$= 8\tfrac{3}{4} \text{ m}$$

Example 2 Multiplication and subtraction
The length of a running track is $\frac{2}{5}$ km. If a runner has completed $5\frac{1}{2}$ laps of a 3 km race, how much further has he to go?
The distance is $3 - (5\tfrac{1}{2} \times \tfrac{2}{5}) = 3 - \dfrac{11}{2} \times \dfrac{2}{5}$ (cancel by 2)

$$= 3 - \frac{11}{5}$$

$$= 3 - 2\tfrac{1}{5}$$

$$= \tfrac{4}{5} \text{ km (i.e. 2 laps)}$$

Example 3 Division and multiplication
If $1\frac{4}{5}$ Canadian dollars are equivalent to $1\frac{2}{5}$ Australian dollars, how many Canadian dollars are equivalent to 84 Australian dollars?

$$84 \times 1\tfrac{4}{5} \div 1\tfrac{2}{5} = 84 \times \frac{9}{5} \div \frac{7}{5}$$

$$= 84 \times \frac{9}{5} \times \frac{5}{7} \quad \text{(cancel by 5 and 7)}$$

$$= 12 \times \frac{9}{1} \times \frac{1}{1}$$

$$= 108 \text{ Canadian dollars}$$

Example 4 Multiplication and ratio
A car radiator will not freeze if the ratio of a particular antifreeze to water is at

least 3:7. Find how much antifreeze would be required in a radiator of capacity 4 litres.

The proportion of antifreeze required is 3 parts in 10, therefore the quantity of antifreeze required is

$$\frac{3}{10} \text{ of 4 litres} = \frac{3}{10} \times 4000 \text{ ml}$$

$$= 1200 \text{ ml}$$

1.11 Precedence rules

So far in this section we have considered the four basic operations of addition, subtraction, multiplication and division, plus the laws which apply to these operations. In some cases we have used brackets to enclose a particular operation which should be done first before the rest of the calculation. You may find it helpful to memorize the order of precedence of these operations in the following form:

Brackets
Of
Division
Multiplication
Addition
Subtraction

Remembering 'BODMAS' will make it easy for you to keep to the right order when simplifying a complex expression.

Wherever possible we use the ordinary curved brackets ( ), but this can be confusing if we have to put one set of brackets *inside* another. In such cases we also use square brackets [ ], and it is usual to have curved brackets inside square ones rather than the other way round. The operation enclosed by the innermost brackets should be done first.

Example $12 + 3[10 - 2(2\frac{4}{5} \times 1\frac{1}{4})]$

The first stage is to simplify the section within the inner curved brackets:

$12 + 3[10 - 2(\frac{14}{5} \times \frac{5}{4})]$ (cancel by 2 and 5)

$12 + 3[10 - 2(\frac{7}{1} \times \frac{1}{2})]$

$12 + 3[10 - 2(\frac{7}{2})]$

We note that $2(\frac{7}{2})$ is equivalent to $2 \times \frac{7}{2}$, which gives

$12 + 3[10 - 2 \times \frac{7}{2}]$

The next stage is to simplify the section within the square brackets, and we note that our precedence rules require us to multiply before we subtract.

$12 + 3[10 - 7]$

$12 + 3[3]$

$12 + 3 \times 3$

$$\frac{12 + 9}{21}$$

It is not always realized that the presence of brackets is implied in some expressions by the use of a horizontal line. Thus

$\dfrac{a + b}{c}$ is equivalent to $(a + b)/c$, i.e. $(a + b) \div c$

$\dfrac{a}{b + c}$ is $a/(b + c)$, i.e. $a \div (b + c)$

$\sqrt{a + b}$ is $\sqrt{(a + b)}$

Example Simplify

$$\frac{2\frac{1}{4}}{4\frac{2}{5} - 3\frac{3}{20}} - \frac{2\frac{1}{2} + 1\frac{2}{5}}{3\frac{1}{4}}$$

In this example the presence of the line above $4\frac{2}{5} - 3\frac{3}{20}$ has the same effect as a bracket and gives priority to this operation. The same applies to the line under $2\frac{1}{2} + 1\frac{2}{5}$.

$$\frac{2\frac{1}{4}}{4\frac{2}{5} - 3\frac{3}{20}} - \frac{2\frac{1}{2} + 1\frac{2}{5}}{3\frac{1}{4}} = \frac{2\frac{1}{4}}{1\frac{8-3}{20}} - \frac{3\frac{5+4}{10}}{3\frac{1}{4}}$$

$$= 2\frac{1}{4} \div 1\frac{5}{20} - 3\frac{9}{10} \div 3\frac{1}{4}$$

$$= 2\frac{1}{4} \div 1\frac{1}{4} - 3\frac{9}{10} \div 3\frac{1}{4} \quad \text{(cancel by 5)}$$

$$= \frac{9}{4} \div \frac{5}{4} - \frac{39}{10} \div \frac{13}{4}$$

$$= \frac{9}{4} \times \frac{4}{5} - \frac{39}{10} \times \frac{4}{13} \quad \text{(cancel by 4, 2, 13)}$$

$$= \frac{9}{1} \times \frac{1}{5} - \frac{3}{5} \times \frac{2}{1}$$

$$= \frac{9}{5} - \frac{6}{5}$$

$$= \frac{3}{5}$$

Exercise A3
1. What fraction of £1 is 75p?
2. Express 60 cents as a fraction of $1.
3. Convert 625 ml to a fraction of a litre.
4. What fraction of a day is 80 minutes?
5. Simplify (a) $3\frac{3}{4} + 2\frac{1}{2}$, (b) $3\frac{3}{4} - 2\frac{1}{2}$.
6. Evaluate (a) $4\frac{2}{5} + 2\frac{7}{10}$, (b) $4\frac{2}{5} - 2\frac{7}{10}$.
7. Simplify (a) $1\frac{2}{3} \times 1\frac{4}{5}$, (b) $1\frac{1}{6} \times 1\frac{1}{7}$.
8. Reduce to its lowest terms $2\frac{2}{5} \times 1\frac{5}{7} \times 2\frac{1}{16}$.
9. Find the value of (a) $\frac{3}{4}$ of $5\frac{1}{3}$, (b) $\frac{5}{8}$ of $2\frac{2}{5}$.
10. Calculate (a) $\frac{7}{8}$ of a metre, (b) $\frac{4}{9}$ of a day.
11. Simplify (a) $4\frac{9}{10} \div 1\frac{2}{3}$, (b) $5\frac{5}{8} \div 2\frac{1}{3}$.
12. In nitric acid the proportions of the elements H:N:O by weight are $1:14:48$. What fraction is nitrogen?

13. Given that the relative atomic masses of hydrogen, oxygen and sulphur are in the ratio $1:16:32$, find the fraction by mass of sulphur in the gases H_2S and SO_2.

14. If a storage tank is holding 360 litres when it is three-quarters full, how much will it contain when it is two-thirds full?

15. Two places are 48 mm apart on a map. If the map is drawn to a scale of $1:25\,000$, find the actual distance between the two places.

16. A carpet salesman starts with a 40 m roll of carpet and cuts off lengths of $4\frac{1}{4}$ m, $3\frac{1}{2}$ m, $5\frac{1}{3}$ m, $3\frac{3}{4}$ m, $4\frac{2}{3}$ m and $3\frac{1}{4}$ m. What length is left?

Simplify as far as possible:

17. $2\frac{2}{3} - 1\frac{1}{2} + 3\frac{1}{4}$

18. $3\frac{1}{3} \times 2\frac{3}{5} - 4\frac{2}{3}$

19. $6\frac{7}{8} \times 3\frac{1}{5} \div 2\frac{3}{4}$

20. $2\frac{1}{2} + \frac{1}{3}$ of $2\frac{1}{4}$

21. $(7\frac{1}{2} - 5\frac{5}{6} \times (6\frac{2}{5} - 4\frac{3}{10})$

22. $8\frac{1}{3} + 2[3 - 4(1\frac{3}{4} - 1\frac{1}{8})]$

23. $\dfrac{8}{5 - 3\frac{1}{2}} - \dfrac{3}{2 - 1\frac{1}{3}}$

24. $\dfrac{1}{4\frac{4}{5} - 2\frac{7}{10}} - \dfrac{5\frac{1}{2} - 2\frac{1}{4}}{7}$

25. $\frac{1}{4}(9\frac{2}{3} - 5\frac{1}{3}) + \frac{2}{3}(9\frac{3}{5} - 3\frac{3}{4})$

2 Decimals and basic arithmetical operations

2.1 Decimal fractions

We write 2835 to represent 2 thousands + 8 hundreds + 3 tens + 5 units. Note the importance of place value, in that moving one place at a time to the right through the number we go from thousands → hundreds → tens → units, i.e. each place moved to the right gives a value only one-tenth as large. This process can be continued to give fractional parts in the sequence tenths, hundredths, thousandths, etc. To distinguish between the whole number and the fractional parts we place a *decimal point* between the units and the tenths, e.g. 26.749 represents 2 tens + 6 units + 7 tenths + 4 hundredths + 9 thousandths. This is normally read as 'twenty six point seven four nine'. (It would be misleading to read it as 'twenty six point seven hundred and forty nine', since we have seven tenths, not seven hundred, etc.)

If the decimal number is purely fractional, we put a zero in front of the decimal point to indicate that there is no whole-number part. Thus 0.638 represents six tenths + three hundredths + eight thousandths.

2.2 Converting decimals to fractions

Since $\quad 0.638 = \dfrac{6}{10} + \dfrac{3}{100} + \dfrac{8}{1000} = \dfrac{638}{1000}$

which can be cancelled to give $\frac{319}{500}$ in its simplest form, we see that a decimal can be put as a common fraction. It is necessary to use a denominator with as many zeros after the 1 as there are figures after the decimal point:

12

e.g.　　$0.1257 = \dfrac{1257}{10\,000}$　　　　$0.075 = \dfrac{75}{1000} = \dfrac{3}{40}$

After converting a decimal to a fraction, the result should be simplified as far as possible:

e.g.　　$0.016 = \dfrac{16}{1000} = \dfrac{2}{125}$　and　$2.375 = \dfrac{2375}{1000} = 2\tfrac{3}{8}$

For the reverse process of converting a fraction to a decimal, we divide the numerator by the denominator. Alternatively, we can multiply the top and bottom of the fraction by a number which changes the denominator to a power of 10:

e.g.　　$\dfrac{1}{4} = 1 \div 4 = 0.25$　or　$\dfrac{1}{4} = \dfrac{25}{100} = 0.25$

$\dfrac{13}{25} = 13 \div 25 = 0.52$　or　$\dfrac{13}{25} = \dfrac{52}{100} = 0.52$

$\dfrac{19}{160} = 19 \div 160 = 0.11875$　or　$\dfrac{19}{160} = \dfrac{19 \times 625}{160 \times 625} = \dfrac{11\,875}{100\,000} = 0.11875$

2.3 Recurring decimals

Some fractions will not cancel or divide out exactly:

e.g.　$\tfrac{1}{3} = 0.3333\ldots$　　　　abbreviated to $0.\dot{3}$

$\tfrac{5}{9} = 0.5555\ldots$　　　　abbreviated to $0.\dot{5}$

$\tfrac{4}{11} = 0.3636\ldots$　　　　abbreviated to $0.\dot{3}\dot{6}$

$\tfrac{7}{13} = 0.538461\,538461\ldots$　　　abbreviated to $0.\dot{5}3846\dot{1}$

$\tfrac{19}{35} = 0.5428571\,428571\ldots$　　　abbreviated to $0.54\dot{2}8\,57\dot{1}$

Note that the section which recurs is signified by the dots placed over the first and last of the figures which repeat.

A decimal fraction which can be expressed exactly in a limited number of decimal places is called a *terminating* decimal, e.g. $\tfrac{3}{8}$ as a decimal is exactly 0.375. Some decimals do not terminate and they either recur or continue without recurrence. The value of π has been worked out to 10\,000 decimal places without any sign of recurrence or termination.

2.4 Decimal places

For practical purposes it is usually necessary to restrict the number of figures after the decimal point, and one effective way of doing this is to stipulate that a decimal fraction should be expressed correct to a given number of decimal places.

Example　Express $\tfrac{1}{3}$ and $\tfrac{2}{3}$ as decimals correct to 4 decimal places.

$\tfrac{1}{3} = 0.3333$　　　$\tfrac{2}{3} = 0.6667$

13

Note that the last figure is rounded up if it is followed by a 6, 7, 8 or 9 – if it is followed by a 5 it is rounded either up or down as necessary to make it an even number.

Examples   0.2468 correct to 3 decimal places is 0.247
2.5432 correct to 2 decimal places is 2.54
0.0775 correct to 3 decimal places is 0.078
0.4545 correct to 2 decimal places is 0.45
0.0208 correct to 3 decimal places is 0.021
0.403 correct to 2 decimal places is 0.40

2.5 Significant figures
All the figures in a number are significant except for zeros placed there to give the position of the decimal point.

Examples   5.4321 correct to 3 significant figures is 5.43
0.456 correct to 2 significant figures is 0.46
135.7 correct to 2 significant figures is 140
0.006 66 correct to 2 significant figures is 0.0067
10.8102 correct to 5 significant figures is 10.810

Note in the last of these examples that the final zero is significant. 10.810 is correct to 5 significant figures (3 decimal places); 10.81 is only correct to 4 figures and could have been the result of reducing to 4 significant figures any value between 10.8051 and 10.8149.

2.6 Converting to percentages
Percentages are fractions with a denominator of 100. Any fraction with a denominator which is a factor of 100 is therefore easily converted into a percentage.

Examples $\frac{2}{5} = \frac{40}{100} = 40\%$

$\frac{3}{4} = \frac{75}{100} = 75\%$

The general rule is to multiply the fraction by 100% and cancel wherever possible.

Examples $\frac{4}{25} \times 100\% = \frac{4}{1} \times 4\% = 16\%$

$\frac{3}{8} \times 100\% = \frac{3}{2} \times 25\% = \frac{75}{2}\% = 37\frac{1}{2}\%$

With decimals the same general rule applies, but it should be noted that the apparent effect of multiplying a decimal by 100 is a shift of the decimal point two places to the right.

Examples $0.423 \times 100\% = 42.3\%$

$0.018 \times 100\% = 1.8\%$

Percentages can be converted to fractions or decimals by the reverse process.

Examples $45\% = \dfrac{45}{100} = 0.45$ or $\frac{9}{20}$

$32\% = \dfrac{32}{100} = 0.32$ or $\frac{8}{25}$

2.7 Addition and subtraction of decimals
To avoid mistakes, it is always best to write the numbers to be added (or subtracted) in vertical order with the decimal points in line.

Examples
a) Add 16.210, 3.407 and 0.095.
   This is written down as

   16.210
   3.407
   +0.095
   ‾‾‾‾‾‾‾
   19.712

b) Subtract 0.927 from 3.405.
   3.405
   −0.927
   ‾‾‾‾‾‾
   2.478

2.8 Multiplication and division of decimals
There are four cases to be considered.

a) Multiplying a decimal by an integer, e.g. 0.469×7.
   0.469
   7
   ‾‾‾‾‾
   3.283

Note that there are three decimal places in 0.469, so there should be three decimal places in the answer (3.283).

b) Multiplying a decimal by another decimal, e.g. 4.05×0.72.

This is $\dfrac{405}{100} \times \dfrac{72}{100}$

$405 \times 72 = 29\,160$

and $\quad 100 \times 100 = 10\,000$

$\therefore \quad \dfrac{405}{100} \times \dfrac{72}{100} = \dfrac{29\,160}{10\,000} = 2.9160$

15

The two decimals being multiplied together each have 2 decimal places and this gives 4 decimal places in the product. (See the note on page 25 about restricting the number of decimal places in answers to practical problems.)

c) Dividing a decimal by an integer, e.g. $0.469 \div 7$.

$$7 \underline{|0.469}$$
$$\underline{0.067}$$

In this case, the answer will only have the same number of decimal places as the original decimal if the integer divides into it exactly, as in the example above. If it does not divide exactly, then the answer should be limited to a realistic number of decimal places.

e.g. $4.528 \div 13$ by long division:

$$
\begin{array}{r}
0.348\,307\ldots \\
13\overline{)\,4.528\,000} \\
\underline{39} \\
62 \\
\underline{52} \\
108 \\
\underline{104} \\
40 \\
\underline{39} \\
100 \\
\underline{91} \\
9
\end{array}
$$

The original decimal (4.528) was given to 4 significant figures and the result of the division should therefore not be given to more than 4 significant figures and 3 significant figures would usually be sufficient if this was the final answer to a calculation. Hence we would write the answer as 0.348 or 0.3483.

d) Dividing a decimal by another decimal, e.g. $4.05 \div 0.72$.

This is $\dfrac{4.05}{0.72}$ and we can remove the decimal point by multiplying numerator and denominator (top and bottom) by 100 to give

$$\frac{405}{72} = 5.625 \quad \text{(by long division)}$$

Similarly, $4.05 \div 0.072$ would be

$$\frac{4.05}{0.072} = \frac{4050}{72} = 56.25$$

Exercise A4
1. Write three tenths, six hundredths as a single decimal.
2. Convert the following decimals to fractions:
   (a) 0.6   (b) 0.25   (c) 0.625
3. Put each of the following decimals into fractional form, simplifying the

16

result as far as possible:
(a) 0.024   (b) 0.0875   (c) 1.025
4. Convert the following fractions to decimals:
(a) $\frac{4}{5}$ (b) $\frac{13}{16}$ (c) $\frac{11}{25}$
5. Put each of the following fractions into decimal form:
(a) $\frac{1}{8}$ (b) $\frac{11}{32}$ (c) $\frac{9}{125}$
6. Write down the recurring decimals corresponding to the following fractions:
(a) $\frac{7}{9}$ (b) $\frac{8}{11}$ (c) $\frac{4}{7}$
7. Express each of the following fractions in the form of a recurring decimal:
(a) $\frac{9}{11}$ (b) $\frac{9}{13}$ (c) $\frac{9}{14}$
8. Which of the following fractions would be a recurring decimal if converted?
$\frac{1}{2}, \frac{1}{3}, \frac{1}{4}, \frac{1}{5}, \frac{1}{6}, \frac{1}{7}, \frac{1}{8}, \frac{1}{9}, \frac{1}{10}, \frac{1}{11}, \frac{1}{12}, \frac{1}{13}, \frac{1}{14}, \frac{1}{15}, \frac{1}{16}$
9. Which is greater, $\frac{5}{9}$ or 0.55?
10. Which is greater, $3\frac{1}{7}$ or 3.142?
11. Express each of the following numbers correct to 2 decimal places:
(a) 3.141 59   (b) 0.0709   (c) 4.545
12. Express each of the following numbers correct to 2 significant figures:
(a) 63.92   (b) 0.0709   (c) 0.0702
13. Add together 4.73, 0.19, 2.05.
14. Add together 21.3, 4.38, 1.82.
15. Subtract 2.153 from 9.628.
16. From 13.261 take 8.397.
17. Simplify 12.48 + 6.21 − 8.37.
18. Find the length of wire left if 12.85 metres are cut from a 20 metre coil.
19. Find the values of (a) 1.392 × 9, (b) 0.0576 × 7.
20. Evaluate (a) 4.196 × 17, (b) 0.093 × 22.
21. Divide 2.848 by 4.
22. Divide 0.620 by 8.
23. Multiply 4.25 by 3.16.
24. Divide 32.04 by 4.5.
25. Evaluate $\dfrac{2.07 \times 10.24}{7.2}$

3 Indices

3.1 Definitions

The idea of factors was introduced in section 1.1 and we saw that some numbers had repeated factors, e.g. $12 = 2 \times 2 \times 3$. When a factor is repeated many times, e.g. $32 = 2 \times 2 \times 2 \times 2 \times 2$, we can use a different way of writing this down and so avoid the long string of repeated factors. Using what is called *index notation*, we can write $32 = 2^5$, where 2^5 implies a string of five twos multiplied together, i.e. $2^5 = 2 \times 2 \times 2 \times 2 \times 2$.

In this notation, the lower number (in large type) is known as the *base* and the upper number (usually printed in smaller type) is called the *index*. The plural of index is *indices*. 2^5 has base 2 and index 5. It is usually read as 'two to the *power* five', or more correctly as 'two raised to the power of five'. Whole-number values of the index are commonly referred to as *powers* in this way.

17

The most common powers have special names, so that 5^2 would be referred to as 'five *squared*' and 5^3 would be 'five *cubed*'.

When the index is fractional we have *roots*, but we only need to consider the simplest case, i.e. when the index is $\frac{1}{2}$ we have what are called *square roots*.

When the index is −1, we have a *reciprocal*, defined as 1 divided by the base.

Examples

3^4 has base 3 and index 4. It means $3 \times 3 \times 3 \times 3$. It is read as 'three to the power four'.

$7^{\frac{1}{2}}$ has base 7 and index $\frac{1}{2}$. It means $\sqrt{7}$. It is read as 'the square root of seven'.

6^{-1} has base 6 and index −1. It means $\frac{1}{6}$. It is read as 'the reciprocal of six'.

Squares, square roots and reciprocals of numbers can all be found by using mathematical tables (see section 6).

Index notation can also be used in algebra, where letters are used in place of some of the numbers. It will help us in considering the laws of indices if we use the letter a to represent any base and letters m and n to represent indices. Thus we write a^2 for $a \times a$ and $a^{\frac{1}{2}}$ or $\sqrt{a}$ for the square root of a, so that $\sqrt{a} \times \sqrt{a} = a$. Similarly a^n implies 'a raised to the power of n' and a^{-1} means the reciprocal, $\frac{1}{a}$.

3.2 Laws of indices

Remembering that $2^2 = 2 \times 2$ and $2^3 = 2 \times 2 \times 2$, we see that

$$2^2 \times 2^3 = 2 \times 2 \times 2 \times 2 \times 2 = 32 \text{ or } 2^5$$

i.e. $2^2 \times 2^3 = 2^{2+3} = 2^5$

This principle can be stated as a law as follows:

to multiply powers of the same base, add the indices.

In its most general form we write this as

$$a^m \times a^n = a^{m+n}$$

It is easy to see that there must be a corresponding law for division:

to divide powers of the same base, subtract the indices.

This can be written as

$$a^m \div a^n = a^{m-n}$$

and care must be taken to see that the index of the second is subtracted from the index of the first.

It could also be written as

$$\frac{a^m}{a^n} = a^{m-n}$$

and it is clear that the index of the denominator must be subtracted from the index of the numerator (i.e. top index minus bottom index):

e.g. $\dfrac{2^3}{2^2} = 2^{3-2} = 2^1 = 2$

The last of the three rules for indices concerns the case when an expression consisting of a base raised to a power is itself raised to a power:

to raise to a power, multiply the indices.

This means $(a^m)^n = a^{mn}$

Examples $x^5 \times x^3 = x^8$ $2x^5 \times 4x^3 = 8x^8$

$x^5 \div x^3 = x^2$ $2x^5 \div 4x^3 = \frac{1}{2}x^2$

$(x^5)^3 = x^{15}$ $(2x^5)^3 = 8x^{15}$

3.3 Special cases

1. When $m = n$, the three rules give us the following results:

(i) $a^n \times a^n = a^{2n}$ (indices can contain letters *and* numbers)

(ii) $a^n \div a^n = a^0$ $\left(\dfrac{a^n}{a^n} = 1 \text{ by cancelling, therefore } a^0 = 1 \right)$

(iii) $(a^n)^n = a^{n^2}$ (an index can be raised to a power)

2. When $m = 0$, the relationship

$$\frac{a^m}{a^n} = a^{m-n} \qquad \text{becomes} \qquad \frac{1}{a^n} = a^{-n}$$

and this illustrates that a negative index corresponds to a reciprocal.

e.g. $x^{-1} = \dfrac{1}{x}$ $x^{-2} = \dfrac{1}{x^2}$

but note that $2x^{-3} = 2(x^{-3}) = 2\left(\dfrac{1}{x^3}\right) = \dfrac{2}{x^3}$

3. When $n = \frac{1}{2}$, the relationship

$$a^n \times a^n = a^{2n} \qquad \text{becomes} \qquad a^{\frac{1}{2}} \times a^{\frac{1}{2}} = a^1 = a$$

Since $\sqrt{a} \times \sqrt{a} = a$, it follows that $a^{\frac{1}{2}} = \sqrt{a}$, i.e. $a^{\frac{1}{2}}$ is the square root of a.

e.g. $4^{\frac{1}{2}} = \pm 2$

$x^{1.5} = x^1 \times x^{0.5} = x\sqrt{x}$

Expressions involving indices

Care must be taken with expressions involving numbers, letters, roots and indices to ensure that each expression is written in a way which makes clear exactly what is intended.

Compare $3x^{\frac{1}{2}}$, $(3x)^{\frac{1}{2}}$ and $3^{\frac{1}{2}}x$.

$3x^{\frac{1}{2}}$ is $3\sqrt{x}$

$(3x)^{\frac{1}{2}}$ is $\sqrt{3x}$

$3^{\frac{1}{2}}x$ is $\sqrt{3}x$

Compare $\dfrac{8}{x^3}$ and $\dfrac{1}{8x^3}$.

$\dfrac{8}{x^3}$ is $\left|8x^{-3}\right.$

$\dfrac{1}{8x^3}$ is $\dfrac{1}{8}x^{-3}$ $\qquad (2x)^{-3}$

To combine two expressions involving indices, it is necessary to ensure that they have a common base.

Examples $2^{3x} \times 8^x = 2^{3x} \times 2^{3x} = 2^{6x}$

$2^{3x} \div 8^x = 2^{3x} \div 2^{3x} = 1$

$2^{3x} + 8^x = 2^{3x} + 2^{3x} = 2 \times 2^{3x} = 2^{3x+1}$

$(3x)^2 + 3x^2 = 9x^2 + 3x^2 = 12x^2$

$(3x)^2 \div 3x^2 = 9x^2 \div 3x^2 = 3$

$\sqrt{4x^3} \times 2\sqrt{x} = 2x^{\frac{3}{2}} \times 2x^{\frac{1}{2}} = 4x^2$

$\sqrt{4x^3} \div 2\sqrt{x} = 2x^{\frac{3}{2}} \div 2x^{\frac{1}{2}} = x$

$\sqrt{4x^3} + 2\sqrt{x} = 2x\sqrt{x} + 2\sqrt{x} = 2\sqrt{x}(x+1)$

Exercise A5
1. Evaluate (a) 2^4, (b) 3^3, (c) 10^3, (d) $9^{\frac{1}{2}}$.
2. Evaluate (a) 4^0, (b) 4^1, (c) $4^{\frac{1}{2}}$, (d) 4^{-1}.
3. Evaluate (a) $9^{1.5}$, (b) $4^{-\frac{1}{2}}$, (c) $(0.25)^{\frac{1}{2}}$, (d) $(0.16)^{-0.5}$.
4. Simplify as far as possible each of the following:

    (a) $\left(\dfrac{4}{x^2}\right)^{0.5}$ (b) $\left(\dfrac{1}{9x^4}\right)^{\frac{1}{2}}$ (c) $\left(\dfrac{1}{2x^3}\right)^{-1}$ (d) $\left(\dfrac{9}{4x^2}\right)^{-0.5}$

5. Simplify (a) $25^x \times 5^x$, (b) $25^x \div 5^x$, (c) $25^x - 5^x$.
6. Simplify $3^{4x+1} \div 9^{2x}$.
7. Simplify $6^{\frac{1}{2}} \times 2^{1\frac{1}{2}} \times 3^{2\frac{1}{2}}$.
8. Simplify $(4a^2 b^4)^{0.5}$.
9. Evaluate pv^n when $p = 760$, $v = 4$, $n = 1.5$.
10. Simplify (a) $\sqrt{9x^2} + \sqrt{4x^2}$, (b) $\sqrt{27x^3} \times \sqrt{3x}$.

4 Numbers in standard form and in binary form

4.1 Expressing a number in standard form

If we take the distance to the sun as being 150 000 000 000 m we have a rather unwieldy number, but we then find that distances to the stars are many thousands of times larger! Since it is inconvenient to have numbers with so many zeros, it is preferable to use index notation. A number in standard form will be in the form $a \times 10^n$ such that $1 \leqslant a < 10$ and n is an integer, i.e. we require a to be a whole number not less than 1 but less than 10. In this form we would write the distance to the sun as 1.5×10^{11} m.

20

Standard form can be used for very small numbers as well as very large ones. For example, the wavelength of violet light is 0.000 000 4 m and we can write this as 4×10^{-7} m.

Example Express the following numbers in standard form: 41 500, 0.0072, 380 000 000, 0.000 035.

$$41 500 = 4.15 \times 10^4$$

$$0.0072 = 7.2 \times 10^{-3}$$

$$380 000 000 = 3.8 \times 10^8$$

$$0.000 035 = 3.5 \times 10^{-5}$$

4.2 Conversion to decimal form
Given a number in standard form, such as 1.234×10^5, we can easily convert this to normal decimal form. All that is necessary is to find the right position for the decimal point.

e.g. 1.234×10^5 is $1.234 \times 10 \times 10 \times 10 \times 10 \times 10$

$$= 12.34 \times 10 \times 10 \times 10 \times 10$$

$$= 123.4 \times 10 \times 10 \times 10$$

$$= 1234 \times 10 \times 10$$

$$= 12 340 \times 10$$

$$= 123 400$$

Although 10^5 means multiplying by 10 five times, it was not actually necessary to do it in stages for we see that each multiplication by 10 simply involved a shift of the decimal point one place further to the right. Multiplying 1.234 by 10^5 can be done simply by moving the decimal point five places in the positive direction (note that we insert any extra zeros which may be necessary):

$$1 . 2 3 4 0 0$$

If the index is negative, the decimal point is moved in the opposite direction, so 6.54×10^{-3} becomes

$$0 . 0 0 6 5 4$$

4.3 Operations on numbers in standard form
For *addition* or *subtraction*, the numbers must have the same power of 10. Thus 2.3×10^5 may be added to 1.2×10^5 to give 3.5×10^5, but to add 3.2×10^5 to 4×10^4 is not so straightforward. The easiest way is to change 4×10^4 to 0.4×10^5 and *then* we can add it to 3.2×10^5 to give 3.6×10^5. If we add 8×10^3 to 9×10^3 we get 17×10^3 and this is no longer in standard form since 17 is not between 1 and 10. In this case we change 17×10^3 to 1.7×10^4. Similarly, if we subtract 1.6×10^7 from 2.3×10^7 we are left with 0.7×10^7 which has to be changed to 7.0×10^6.

For *multiplication* or *division* it is usual to use logarithms by the method shown in section 6.4, but for some simple cases that is not necessary.

Example   Evaluate (a) $5 \times 1.6 \times 10^3$, (b) $\dfrac{6 \times 10^7}{4 \times 10^5}$.

a) $5 \times 1.6 \times 10^3 = 8 \times 10^3$

b) $\dfrac{6 \times 10^7}{4 \times 10^5} = (6 \div 4) \times 10^{7-5} = 1.5 \times 10^2$

Exercise A6
1. Express the following numbers in standard form:
   375 000   2173   0.015   0.000 125
2. Convert into ordinary numbers
   3.2×10^3 4×10^6 8.3×10^{-4} 6.91×10^{-2}
3. Simplify the following expressions, giving the answers in standard form:
   $4.3 \times 10^5 + 2.8 \times 10^5$ $7 \times 10^{-2} + 6 \times 10^{-3}$

4. Simplify (a) $\dfrac{2.1 \times 10^6}{7 \times 10^3}$, (b) $\dfrac{6 \times 10^{-5}}{3 \times 10^{-7}}$.

5. If the distance to a certain star is 3×10^{38} m, what would be the distance to a star twice as far away?

4.4 Numbers in binary form
The number system we normally use has a base of ten and we call it the decimal or *denary* system. As we saw in section 2.1, we write down numbers in such a way that moving a figure one place to the left gives it a value *ten* times greater, so that, from the decimal point, we move through units to tens, hundreds, thousands, etc.

There is another number system it is convenient to use with computers and punched-card systems. This has a base of two and is known as the *binary* system. Instead of using the figures 0 to 9, the binary system uses only 0 to 1. In this system, moving a figure one place to the left gives it a value *two* times greater, so that, from the unit position, we successively double the place value.

To convert a decimal number to binary form, we simply divide successively by two and note the remainders at each stage.

Example   Convert 43 to binary form.

    2 |43
    2 |21 remainder 1
    2 |10 remainder 1
    2 | 5 remainder 0
    2 | 2 remainder 1
    2 | 1 remainder 0
       0 remainder 1

Therefore 43 is 101011 in binary form (from the remainders, reading upwards).

4.5 Converting binary to denary numbers

We have seen that, in the binary system, place value doubles as we move to the left. For numbers less than 64 we need up to six figures and the place values will be from 2^0 to 2^5. In denary equivalents, $2^0 = 1$, $2^1 = 2$, $2^2 = 4$, $2^3 = 8$, $2^4 = 16$, $2^5 = 32$. Using these equivalents, we should be able to convert 101011 back to 43.

Writing

$$2^5 \ 2^4 \ 2^3 \ 2^2 \ 2^1 \ 2^0$$
$$1 \ \ 0 \ \ 1 \ \ 0 \ \ 1 \ \ 1$$

we see that

$$101011 = 1 \times 2^5 + 0 \times 2^4 + 1 \times 2^3 + 0 \times 2^2 + 1 \times 2^1 + 1 \times 2^0$$
$$= \ 32 \ + \ 0 \ + \ 8 \ + \ 0 \ + \ 2 \ + \ 1$$
$$= \ 43$$

With practice it becomes easy to miss out the line giving the powers of 2, so that we could convert the binary number 111101 as follows:

$$111101 = 32 + 16 + 8 + 4 + 0 + 1 = 61$$

Here are some more examples:

$$1011 = 8 + 0 + 2 + 1 = 11$$
$$10001 = 16 + 0 + 0 + 0 + 1 = 17$$
$$101000 = 32 + 0 + 8 + 0 + 0 + 0 = 40$$
$$111001 = 32 + 16 + 8 + 0 + 0 + 1 = 57$$

4.6 Adding binary numbers

Look again at the last of the examples given above. Can you see that 111001 is given by adding together the two previous binary numbers? 10001 + 101000 = 111001 (as we would expect since 17 + 40 = 57).

To avoid making mistakes it is advisable to write binary numbers in vertical order when adding them.

Example　　Add together 100100, 10010, 1001.

```
100100
 10010
  1001
_____
111111
```

(You may like to check that this is equivalent to 36 + 18 + 9 = 63.)

Obviously that was a particularly simple example since the total in any column did not exceed 1. Whenever it does, we have to carry over into the next column, remembering that whenever we get up to 2 that this is equivalent to another unit to be carried into the next column.

Example   Add together 10110, 10101.

```
  10110
  10101
─────────
 101011
```

In this example in both the 2^4 and 2^2 columns we had $1 + 1 = 2$, which carried over as a 1 into the next column leaving a zero behind. If a column adds up to 3, then we leave a 1 as the total in that column and carry over 2 as a 1 in the next column.

Example   Add together 10100, 1101, 1011.

```
  10100
   1101
   1011
─────────
 101100
```

$$2^5 + 2^3 + 2^2$$
$$32 + 8 + 4 = 44$$

(Equivalent to $20 + 13 + 11 = 44$.)

4.7 Uses of the binary system

In any binary number, each digit is either 0 or 1. Since this represents a simple either/or choice between just two alternatives, it can be related to any other system with a similar choice between two alternatives such as 'on/off' or 'yes/no' or 'present/absent'. In practical terms, binary numbers can be entered on punched cards by holes punched in certain marked squares, or computers can mark multiple-choice answers to examination questions by checking whether a particular answer has been selected or not, etc. A whole complex system of operations in computing has been built up by such applications of binary notation.

Exercise A7

1. Put the following numbers into binary form: 3, 9, 29, 37, 58.
2. Convert to denary form 101, 1111, 11000, 110011, 100111.
3. Calculate the answers to the following additions in binary form:
   $1001 + 110$ $1100 + 1111$ $10110 + 11011$
4. Add the following binary numbers and then check your answer by converting each binary number to denary form:
   $11100 + 10001 + 1111$
5. Add 11, 17 and 23 and then put each number into binary form and check that the total is the same.

B   Calculations

5 Reasonable answers to numerical problems

5.1 How many significant figures?

On a level piece of ground, an area 5.9 m by 3.7 m has to be excavated to a depth of half a metre. How much soil has to be removed? Multiplying the three dimensions together will give the volume as 10.915 m³ but in practice there is no virtue in calculating this to five significant figures since it is not helpful, not needed and not reliable! It is not helpful in this case because excavated soil takes up more space than when compressed in the ground where it gets trodden down. It is not needed if the builder only wants to know how many lorry loads it will make. It is not reliable since the initial measurements have only been given to an accuracy of a tenth of a metre and this means that the actual volume could be anywhere between 5.85 x 3.65 x 0.45 and 5.95 x 3.75 x 0.55, i.e. between 9.6 m³ and 12.3 m³. It would be more reasonable therefore to give the soil volume as 11 m³.

Note that our three original dimensions were not all equally accurate, for the measurements of 5.9 m and 3.7 m were given to two significant figures but the depth of 0.5 m was given to one significant figure. In this case we have taken the answer to one significant figure more than the least accurate measurement in the given data, but if all measurements were given to the same number of significant figures we should keep our answer also to this number of significant figures.

One other point to bear in mind concerns the use of approximate values for certain constants. It would be stupid to use a value for π which was less accurate than the measurements of physical dimensions if, for example, we needed to get an accurate value for the volume of a cylinder (see section 18.4).

5.2 Feasible solutions

Certain problems provide two alternative solutions, only one of which is valid. Problems involving finding a maximum or a minimum can sometimes produce two results, one of these corresponding to the maximum and the other to the minimum. It is necessary to select the answer which yields the required condition and reject the other solution. Similarly, in trigonometry it is possible to get multiple solutions when finding an angle by solving an equation. Here again it is necessary to check that you have selected the solution which fits the original conditions.

5.3 Approximate answers

The commonest way of getting a wrong answer to a problem is by making some mistake in the working out! To cover this point, as well as the possible cases mentioned in the last section, it is usually helpful to do a rough estimate of the size of the answer you expect from a particular calculation. This is particularly important when doing calculations by slide rule or with a calculating machine.

Example $\dfrac{2.97 \times 4.05}{6.02 \times 0.97}$

Approximately, this would be $\dfrac{3 \times 4}{6 \times 1} = 2$

more accurately, 2.06 (by calculator).

In this example the numbers were easy to approximate to integers so there was no difficulty in obtaining a reliable estimate. When we have two numbers multiplied together and we wish to simplify the calculation, our approximation will be more accurate if one of the numbers is increased and the other decreased.

Example   4.6 x 6.4

Approximate to 5 x 6 = 30

Accurately, 4.6 x 6.4 = 29.44

5.4 Use of approximations
In some cases, as in the two examples above, the approximate answer is close enough for us to be confident that the answer subsequently obtained by slide rule, calculator or tables is correct. In more complicated cases the approximation may not be good enough for that purpose, but it should still enable us to be sure that we have got the decimal point in the right place. Should the approximation give an answer very different from the final answer, then it will be necessary to check again to find the cause of the error.

6 Using four-figure tables

6.1 Squares, square roots and reciprocals

Squares

We saw in section 3.2 that $a^2 = a \times a$, where a is any number. For small integers it is easy to multiply a number by itself to find the value of a^2, but for large numbers or for decimal fractions it is easier to use a calculator (see section 8) or to use a table of squares. Here is a section of such a table.

Squares (x^2)

x	.0	.1	.2	.3	.4	.5	.6	.7	.8	.9	1	2	3	4	5	6	7	8	9
75	5625	5640	5655	5670	5685	5700	5715	5730	5746	5761	2	3	5	6	8	9	11	12	14
76	5776	5791	5806	5822	5837	5852	5868	5883	5898	5914	2	3	5	6	8	9	11	12	14
77	5929	5944	5960	5975	5991	6006	6022	6037	6053	6068	2	3	5	6	8	9	11	12	14
78	6084	6100	6115	6131	6147	6162	6178	6194	6029	6225	2	3	5	6	8	9	11	13	14
79	6241	6257	6273	6288	6304	6320	6336	6352	6368	6384	2	3	5	6	8	10	11	13	14
80	6400	6416	6432	6448	6464	6480	6496	6512	6529	6545	2	3	5	6	8	10	11	13	14
81	6561	6577	6593	6510	6626	6642	6659	6675	6691	6708	2	3	5	7	8	10	11	13	15
82	6724	6740	6757	6773	6790	6806	6823	6839	6856	6872	2	3	5	7	8	10	12	13	15
83	6889	6906	6922	6939	6956	6972	6989	7006	7022	7039	2	3	5	7	8	10	12	13	15
84	7056	7073	7090	7106	7123	7140	7157	7174	7191	7208	2	3	5	7	8	10	12	14	15

Let us trace through this table the value of $(76.47)^2$. For the first two figures of this number we go down the first column (headed x) until we reach 76. For the third figure we then move across the row horizontally until we find the four-figure number under 0.4. This is 5837 which is the square of 76.4. As we want $(76.47)^2$, we now go across this same row to the 'difference' columns at the right-hand side. Here we find that the difference for the additional 7 on this row is 11. This means that the difference between $(76.40)^2$ and $(76.47)^2$ is 11 which has to be added on to the 5837; thus we get 5848 altogether. Hence $(76.47)^2 = 5848$.

Look at the section of the table of squares again and this time find the value of $(77.83)^2$. Did you get 6058? If not, check again.

Can you deduce the value of $(7.783)^2$? Again we get the four figures 6058, but this time we have to insert a decimal point. One way of doing this is by doing a rough estimate first. Since we see that 7.783 is a little less than 8 and we know that $8^2 = 64$, we infer that $(7.783)^2$ will be a little less than 64. This rough estimate is sufficient to enable us to see that we need to insert the decimal point in the centre of our four figures 6058 to give 60.58. This method of placing the decimal point correctly can also be used when doing calculations by slide rule (see section 7).

Using the same value from the table of squares we would deduce that 778.3^2 would be 605 800, though this value is correct only to four significant figures because the accuracy is limited by the tables we are using. (A calculator would give us the more accurate value of 605 751.)

Without referring to the tables again, you should be able to deduce that $0.7783^2 = 0.6058$.

Square roots

While square roots could be obtained by following through the table of squares in reverse order, it is easier to use a table of square roots. Here is a section from such a table.

Square roots ($\sqrt{x}$)

x	0	1	2	3	4	5	6	7	8	9	1	2	3	4	5	6	7	8	9
78	2793	2795	2796	2798	2800	2802	2804	2805	2807	2809	0	0	1	1	1	1	1	1	2
	8832	8837	8843	8849	8854	8860	8866	8871	8877	8883	1	1	2	2	3	3	4	4	5
79	2811	2812	2814	2816	2818	2820	2821	2823	2825	2827	0	0	1	1	1	1	1	1	2
	8888	8894	8899	8905	8911	8916	8922	8927	8933	8939	1	1	2	2	3	3	4	4	5
80	2828	2830	2832	2834	2835	2837	2839	2841	2843	2844	0	0	1	1	1	1	1	1	2
	8944	8950	8955	8961	8967	8972	8978	8983	8989	8994	1	1	2	2	3	3	4	4	5
81	2846	2848	2850	2851	2853	2855	2857	2858	2860	2862	0	0	1	1	1	1	1	1	2
	9000	9006	9011	9017	9022	9028	9033	9039	9044	9050	1	1	2	2	3	3	4	4	5
82	2864	2865	2867	2869	2871	2872	2874	2876	2877	2879	0	0	1	1	1	1	1	1	2
	9055	9061	9066	9072	9077	9083	9088	9094	9099	9105	1	1	2	2	3	3	4	4	5
83	2881	2883	2884	2886	2888	2890	2891	2893	2895	2897	0	0	1	1	1	1	1	1	2
	9110	9116	9121	9127	9132	9138	9143	9149	9154	9160	1	1	2	2	3	3	4	4	5

You will notice that there are two alternatives to choose from for the square root of any given number. If we select 793 by going down the first column to 79 then across the row to the pair of figures in the 3 column, we see the two alternatives are 2816 and 8905. From the first value we deduce that $\sqrt{7.93}$ = 2.816, $\sqrt{793}$ = 28.16, $\sqrt{79\,300}$ = 281.6, etc. From the second value we get $\sqrt{79.3}$ = 8.905, $\sqrt{7930}$ = 89.05, etc. Having done a rough approximation to the value we expect from a particular square root (as in section 5) we should have no difficulty deciding which of the two alternatives is the correct value to select. For example, if we wish to find $\sqrt{79.3}$, we note that 79.3 is a little less than 81 which is 9^2, so we expect $\sqrt{79.3}$ to be just under 9. The value we want must therefore be 8.905 and we are unlikely to make a mistake by selecting 2.816 or 28.16.

For very large or very small numbers there are two methods we can use to ensure that we choose the correct square-root value and also find the right position for the decimal point. The first method is to use the idea of standard form (see section 4.2). For example, to find the square root of 8 260 000 we write it as 8.26×10^6. The square-root table gives $\sqrt{8.26}$ = 2.874 and $\sqrt{10^6}$ = 10^3, therefore

$$8\,260\,000 = 2.874 \times 10^3 = 2874$$

Similarly, $0.000\,841 = 8.41 \times 10^{-4} = 2.9 \times 10^{-2} = 0.029$

In using this method it is necessary to adjust the number in some cases to give an even power of 10. Thus $0.000\,064 = 6.4 \times 10^{-5}$, but for finding the square root we would write this as 64×10^{-6} and it is then clear that $\sqrt{0.000\,064} = \sqrt{64 \times 10^{-6}} = 8 \times 10^{-3} = 0.008$.

The second method is to separate the number into pairs of digits starting from the decimal point. Thus we write 8 260 000 as 8 26 00 00 and we note that the square root must provide a single digit for each pair in this number:

number   8 26 00 00

square root   2   8   7   4

This method is equally useful for very small numbers:

e.g.   number   0.00 00 64

square root   0. 0   0   8

Reciprocals

The reciprocal of a number is the answer we get when we divide the number into 1. For any number, x, the reciprocal is $\frac{1}{x}$. Four-figure tables of reciprocals have the same layout as corresponding tables of squares etc., but there is one important difference: because the values in the table are becoming smaller as x increases, the figures in the end column need to be *subtracted* not added.

The decimal point can often be placed in the correct position without much difficulty. For example,

$$\frac{1}{1.575} = 0.6349 \quad \text{or} \quad \frac{1}{0.6349} = 1.575$$

For the reciprocals of large or small numbers it may be helpful to use stan-

Reciprocals of numbers ($1/x$)

x	0	1	2	3	4	5	6	7	8	9	Subtract								
											1	2	3	4	5	6	7	8	9
1.0	1.0000	9901	9804	9709	9615	9524	9434	9346	9259	9174	9	18	27	37	46	55	64	73	82
1.1	.9091	9009	8929	8850	8772	8696	8621	8547	8475	8403	8	15	23	30	38	46	53	61	68
1.2	.8333	8264	8197	8130	8065	8000	7937	7874	7813	7752	6	13	19	26	32	38	45	51	58
1.3	.7692	7634	7576	7519	7463	7407	7353	7299	7246	7194	5	11	16	22	27	33	38	44	44
1.4	.7143	7092	7042	6993	6944	6897	6849	6803	6757	6711	5	10	14	19	24	29	33	38	43
1.5	.6667	6623	6579	6536	6494	6452	6410	6369	6329	6289	4	8	13	17	21	25	29	33	38
1.6	.6250	6211	6173	6135	6098	6061	6024	5988	5952	5917	4	7	11	15	18	22	26	29	33
1.7	.5882	5848	5814	5780	5747	5714	5682	5650	5618	5587	3	6	10	13	16	20	23	26	29
1.8	.5556	5525	5495	5464	5435	5405	5376	5348	5319	5291	3	6	9	12	15	17	20	23	26
1.9	.5263	5236	5208	5181	5155	5128	5102	5076	5051	5025	3	5	8	11	13	16	18	21	24

dard form. Thus the reciprocal of 136 200 is found by writing the number as 1.362×10^5 and its reciprocal is then $\frac{1}{1.362} \times 10^{-5}$ which is 0.7342×10^{-5}, i.e. 0.000 007 342. Similarly, to find the reciprocal of 0.000 320 we write this as 3.2×10^{-4} so its reciprocal is $\frac{1}{3.2} \times 10^4$ which is $0.3125 \times 10^4 = 3125$.

Exercise B1
Use four-figure tables of squares, square roots or reciprocals to find the value of each of the following.
1. $3.975^2, 4.732^2, 8.687^2$
2. $43.76^2, 839.1^2, 6921^2$
3. $0.422^2, 0.6834^2, 0.0483^2$
4. $(4.257 \times 10^3)^2, (8.843 \times 10^{-4})^2$
5. $\sqrt{6.843}, \sqrt{5.296}, \sqrt{7.874}$
6. $\sqrt{88.72}, \sqrt{497.3}, \sqrt{3781}$
7. $\sqrt{0.49}, \sqrt{0.049}, \sqrt{0.00049}$
8. $\sqrt{(5.382 \times 10^4)}, \sqrt{(5.382 \times 10^3)}$
9. $\frac{1}{3.46}, \frac{1}{2.896}, \frac{1}{5.953}$
10. $\frac{1}{29.62}, \frac{1}{421.5}, \frac{1}{7325}$
11. $\frac{1}{0.72}, \frac{1}{0.008}, \frac{1}{0.00474}$
12. $\frac{1}{(5.53)^2}, \frac{1}{(0.7429)^2}$
13. $\frac{1}{\sqrt{7.631}}, \frac{1}{\sqrt{0.005478}}$

6.2 Logarithms
In section 3.2 we had the three rules for indices:

$$a^m a^n = a^{m+n} \qquad \frac{a^m}{a^n} = a^{m-n} \qquad (a^m)^n = a^{mn}$$

We can use any convenient value for base a, so for common logarithms we use 10 as our base and the rules become

29

$$10^m \times 10^n = 10^{m+n} \qquad \frac{10^m}{10^n} = 10^{m-n} \qquad (10^m)^n = 10^{mn}$$

As logarithms are indices, our rules tell us that we can multiply two numbers by adding these indices, we can divide by subtracting indices, and we can also find powers — all we need is to be able to express any number as a power of 10. Logarithmic tables enable us to do this.

6.3 Tables of logarithms and antilogarithms

Logarithms ($\log_{10} x$)

x	0	1	2	3	4	5	6	7	8	9	1	2	3	4	5	6	7	8	9
5.0	6990	6998	7007	7016	7024	7033	7042	7050	7059	7067	1	2	3	3	4	5	6	7	8
5.1	7076	7084	7093	7101	7110	7118	7126	7135	7143	7152	1	2	3	3	4	5	6	7	8
5.2	7160	7168	7177	7185	7193	7202	7210	7218	7226	7235	1	2	2	3	4	5	6	7	7
5.3	7243	7251	7259	7267	7275	7284	7292	7300	7308	7316	1	2	2	3	4	5	6	6	7
5.4	7324	7332	7340	7348	7356	7364	7372	7380	7388	7396	1	2	2	3	4	5	6	6	7
5.5	7404	7412	7419	7427	7435	7443	7451	7459	7466	7474	1	2	2	3	4	5	5	6	7
5.6	7482	7490	7497	7505	7513	7520	7528	7536	7543	7551	1	2	2	3	4	5	5	6	7
5.7	7559	7566	7574	7582	7589	7597	7604	7612	7619	7627	1	2	2	3	4	5	5	6	7
5.8	7634	7642	7649	7657	7664	7672	7679	7686	7694	7701	1	1	2	3	4	4	5	6	7
5.9	7709	7716	7723	7731	7738	7745	7752	7760	7767	7774	1	1	2	3	4	4	5	6	7

From the section of a table of logarithms shown above, we see that log 5.14 = 0.7110. This means that $5.14 = 10^{0.711}$. Similarly log 5.65 = 0.7520, which means that $5.65 = 10^{0.752}$. If we wish to find 5.14 x 5.65, our rule tells us that we simply add the two logarithms, so

$$5.14 \times 5.65 = 10^{0.711} \times 10^{0.752} = 10^{1.463}$$

To convert $10^{1.463}$ back to a decimal number we use a table of antilogarithms. Since these tables give the antilogarithms of indices from 0 to 0.9999, the index we want (1.463) is outside the range. In such cases we separate the whole number part (the *characteristic*) from the fractional part (the *mantissa*) of the logarithm, i.e. we write

$$10^{1.463} = 10^1 \times 10^{0.463}$$

We know $10^1 = 10$ and the table of antilogarithms gives us $10^{0.463} = 2.904$, so the complete calculation now becomes

$$5.14 \times 5.65 = 10^{0.711} \times 10^{0.752} = 10^{1.463}$$
$$= 10^1 \times 10^{0.463} = 10 \times 2.904 = 29.04$$

The conventional way of setting out such a calculation is as follows:

No.	Log.
5.14	0.7110
5.65	0.7520
29.04	1.4630

We can find the logarithm of any number, however large, by expressing numbers greater than 10 in standard form (see section 4.2). We have already seen that $10^{0.463} = 2.904$, so log $2.904 = 0.463$. Also we found log 29.04 = 1.463. If we wanted log $2\,904\,000$ we would put $2\,904\,000$ as 2.904×10^6 so that

$$\log 2\,904\,000 = \log 2.904 \times 10^6 = 0.463 + 6 = 6.463$$

Similarly,

$$\log 0.002\,904 = \log 2.904 \times 10^{-3} = 0.463 - 3 = \bar{3}.463.$$

Note in this case that the logarithm has a negative characteristic ($\bar{3}$) but a positive mantissa (.463). Such logarithms are often called bar-logs, because of the bar drawn over the characteristic to indicate that it is negative.

6.4 Multiplication and division by logarithms

As already indicated, two numbers are multiplied by adding their logarithms to find the logarithm of the product. Similarly, subtracting one logarithm from another is the equivalent of division.

Example   Evaluate (a) 172.5×2.048, (b) $172.5 \div 2.048$.

a)

No.	Log.
172.5	2.2368
2.048	0.3113
353.3	2.5481

(by addition)

Answer: 353

b)

No.	Log.
172.5	2.2368
2.048	0.3113
84.24	1.9255

(by subtraction)

Answer: 84.2

6.5 Can a logarithm be zero?

A logarithm is an index and an index can be zero. In section 3.3 we deduced that $a^0 = 1$ where a could be any base. With common logarithms the base is 10. Since this gives us $10^0 = 1$, we deduce that log $1 = 0$.

6.6 Reciprocals

Using logarithms, a reciprocal is simply a special case of division in which a number is divided into 1.

Example   Evaluate $1/6.54$.

No.	Log.
1	0.0000
6.54	0.8156
0.1529	$\bar{1}$.1844

Answer: 0.153

6.7 Powers

From the rule for indices $(a^m)^n = a^{mn}$, we see that raising any number to a

31

positive integral power is done by multiplying the logarithm of the number by the power.

Examples   Evaluate (a) $(6.965)^3$, (b) $(0.9506)^4$.

a)

No.	Log.
6.965	0.8429
	x3
337.8	2.5287

b)

No.	Log.
0.9506	$\bar{1}.9780$
	x4
0.8166	$\bar{1}.9120$

Note that example (b) is a compact way of setting out the following calculation:

$$(0.9506)^4 = (9.506 \times 10^{-1})^4$$
$$= (10^{0.978} \times 10^{-1})^4$$
$$= 10^{3.912} \times 10^{-4}$$
$$= 10^{0.912} \times 10^{-1}$$
$$= 8.166 \times 10^{-1}$$
$$= 0.8166$$

6.8 Square roots

The logarithm of the square root of a number is half the logarithm of the number.

Example   Evaluate $\sqrt{96.43}$.

$$\log 96.43 = 1.9842$$
$$\tfrac{1}{2} \log 96.43 = 0.9921$$

then, taking the antilogarithm,

$$\sqrt{96.43} = 9.819 \qquad \textit{Answer:} \ 9.82$$

The square root of a number less than 1 involves a logarithm with negative characteristic (a bar-log) and this characteristic may need to be made into an even number if it is not already even.

Example   Evaluate 0.001 543.

$$\log 0.001\ 543 = \bar{3}.1844 = \bar{4} + 1.1884$$
$$\tfrac{1}{2} \log 0.001\ 543 = \bar{2} + .5942$$

then, taking the antilogarithm,

$$0.001\ 543 = 0.039\ 28 \qquad \textit{Answer:} \ 0.0393$$

6.9 Calculations using logarithms.

At this level, practical calculations should not involve a combination of more than two of the operations listed above in sections 6.4 to 6.8. The following examples illustrate the types of calculations which might be encountered.

a) 4.65 x 15.64 x 0.327

No.	Log.
4.65	0.6675
15.64	1.1942
0.327	$\bar{1}$.5145
23.78	1.3762

Answer: 23.8

b) $(4.214)^2$ x 835

No.	Log.
4.214	0.6247
	2
	1.2494
835	2.9217
14 830	4.1711

Answer: 14 800

c) $\dfrac{1}{2.721 \times 3.964}$

No.	Log.
2.721	0.4348
3.964	0.5981
	1.0329
0.092 70	$\bar{2}$.9671

Answer: 0.0927

6.10 Accuracy

When using four-figure tables, it is essential to bear in mind that the answers to calculations will be accurate to 3 significant figures but not reliable in the fourth. For this reason, the answers to all the examples in this section have been reduced to 3 significant figures. The general principle is that answers should be restricted to the number of figures that can reasonably be expected to be accurate.

Exercise B2

1. Explain the meaning of the following words: indices, logarithm, base, characteristic, mantissa.
2. Write down the logarithms of the following numbers: 4.567, 456.7, 0.3962, 576 000, 0.030 03.
3. Use a table of antilogarithms to find the numbers corresponding to each of the following logarithms: 0.7654, 0.0897, 1.2345, $\bar{1}$.7359, $\bar{2}$.6007.
4. Use logarithms to evaluate
   (a) 49.14 x 2.018   (b) 1.893 x 0.9654

5. Evaluate by logarithms
   (a) 1.374 x 2.369 x 1.287   (b) 29.62 x 4.397 x 0.0431
6. Use logarithms to divide 463.7 by 2.492.
7. Use logarithms to find the value of
   (a) 29.73 ÷ 4.855   (b) 0.9374 ÷ 0.0566
8. Evaluate $\dfrac{21.47 \times 17.93}{46.62}$.

Further practice may be obtained by using logarithms to work through the questions in exercise B1 on page 29.

7 Using a slide rule

7.1 The slide rule

Figure B1 shows the main parts of a slide rule. The end caps are fixed and they secure the top and bottom stocks rigidly at a fixed distance apart so that the slide moves evenly from one end to the other. The transparent plastic cursor slides between the end caps and is spring-loaded to stay in any position in which it is set. The cursor line enables us to line up one scale with any other on the slide rule.

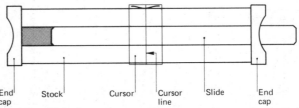

End cap   Stock   Cursor   Cursor line   Slide   End cap

Fig. B1 Parts of a slide rule

The scales
Looking at your slide rule, or the diagrams in fig. B2, you can see that the scales marked A and B are the same. Can you pick out another pair of scales which are identical? Two other identical scales are those marked C and D. These letters appear at the left of the scales, while on the right you will notice that the C and D scales are marked x and the A and B scales correspond to x^2.

All other scales relate to the basic D scale which sets out logarithmically the values of x from 1 to 10.

The following scales may appear on an elementary slide rule:

CF and DF	x	Factor π multiplies C and D scale values.
CI	$\dfrac{1}{x}$	Inverse (i.e. reciprocal) of values on C scale.
DI	$\dfrac{1}{x}$	Inverse of values on D scale.
K	x^3	Cubes of values on C and D scales.
L	$\log x$	Logarithms of values on D scale.

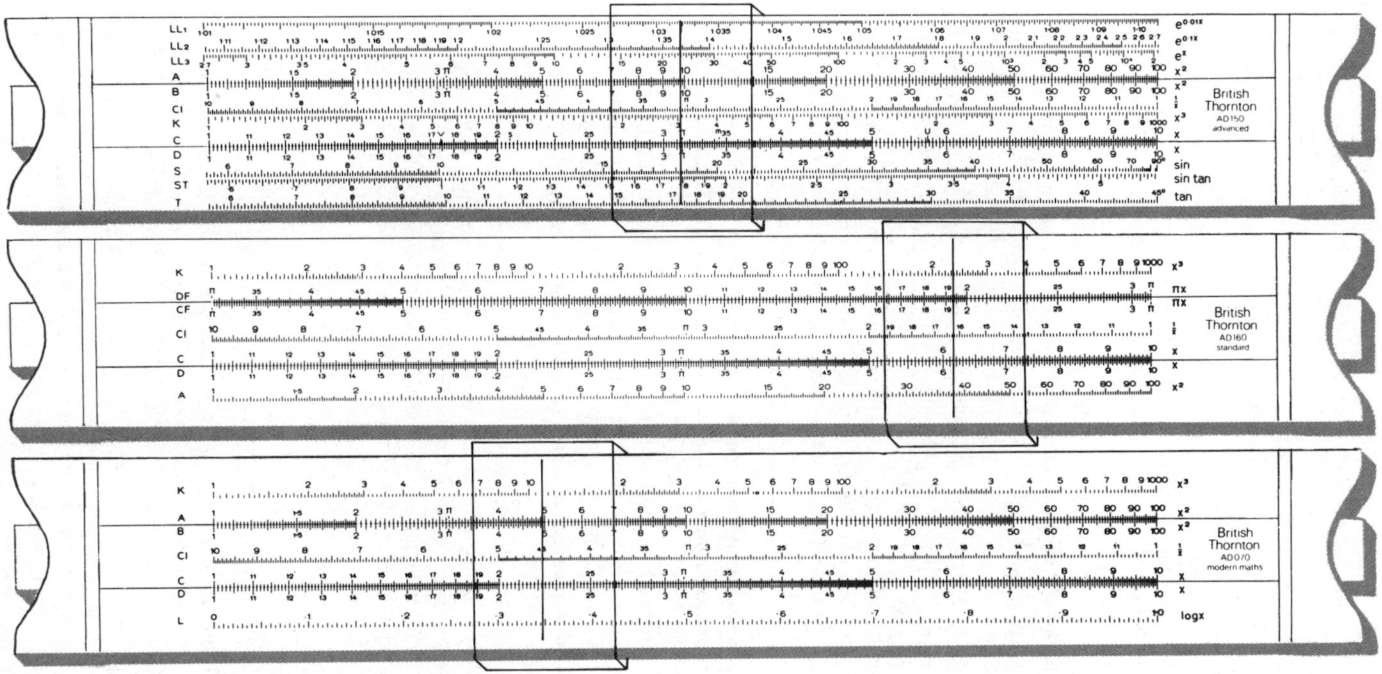

Fig. B2 Slide-rule scales

S	sin ⎫	These scales give values in degrees, the sine or
T	tan ⎭	tangent of which can be read on the D scale.

7.2 Multiplication and division

(You are advised to study the instruction booklet for your slide rule and to practice the examples set out there.)

The procedure for multiplication is illustrated by the following example in which 2.5 is multiplied by 1.8.

i)   Move the slide to bring 1 on the C scale against 2.5 on the D scale.
ii) Move the cursor till the cursor line is over 1.8 on the C scale.
iii) Read the answer (4.5) where the cursor line cuts the D scale.

Multiplication is more accurate on C and D scales, but it is sometimes convenient to use A and B scales instead, especially when multiplying more than two numbers.

To use your slide rule for division, you simply reverse the process used for multiplication. With slide rule in hand, follow through these steps to find the answer when 6.27 is divided by 1.65.

i)   Move the cursor till the cursor line is over 6.27 on the D scale.
ii) Move the slide to bring 1.65 on the C scale under the cursor line.
iii) Read the answer (3.80) on the D scale opposite 1 on the C scale.

7.3 Reciprocals, squares and square roots

The slide rules shown in fig. B2 have a CI scale down the centre of the slide, so reciprocals are found easily and conveniently by moving either the cursor or the slide until the required number on the C scale is directly beneath the cursor line. The reciprocal of this number then appears on the CI scale under the cursor line. In reading the value of this reciprocal, remember that the reciprocal scale should be read in the reverse direction to most other scales or else you may make a mistake in estimating a value which falls between two divisions.

Since A and B scales give x^2 values for the x values of the C and D scales, these scales can be used for calculating squares. It is advisable to match scale A with scale D, using the cursor line to read from a number on the D scale of the lower stock to its square on the A scale of the upper stock. Alternatively you could use the B and C scales on the slide, but try to avoid using one scale on the slide with one on the stock when finding squares or square roots. Practice a few examples on your slide rule and check your results using a table of squares.

The reverse process, going from scale A to scale D (or from B to C) will give the square root of any number between 1 and 100. Practice a few examples and check your results using a table of square roots.

When finding the reciprocal, square or square root of a number outside the range of the appropriate scale on the slide rule, the method of section 6.1 can be used to place the decimal point in the right place. However, it is always wise to estimate an approximate value for the answer as a check and, with practice, it will be found that such a rough estimate is all that is needed to enable the decimal point to be placed correctly.

7.4 Evaluation of expressions

With a complex expression involving both multiplication and division, these operations should be carried out alternately.

Example   Evaluate $\dfrac{3.1 \times 2.7 \times 4.6}{1.9 \times 3.5}$.

With logarithms, we would multiply out the top row first before starting to divide, but with a slide rule we alternate. Thus we adopt the following sequence:
i)   Move the cursor till the cursor line is over 3.1 on the D scale.
ii)   Move the slide to bring 1.9 on the C scale under the cursor line.
iii)   Set the cursor line over 2.7 on the C scale.
iv)   Move the slide again to bring 3.5 on the C scale under the cursor line.
v)   Set the cursor line over 4.6 on the C scale.
vi)   Read the answer on the D scale under the cursor line (5.79).
(Note that the D scale is used only at the start and at the end, with all intermediate steps on the C scale. The cursor is moved for multiplication and the slide for division.)

This example was chosen because all the numbers were between 1 and 10 and could be located on the C and D scales. Similarly, numbers up to 100 can be dealt with using A and B scales. For numbers outside these ranges we can use the methods of section 4 to express the numbers in standard form; then, after performing the calculation, we convert the answer back to normal decimal form.

e.g. $\dfrac{310 \times 0.27 \times 4600}{19 \times 350} = \dfrac{3.1 \times 2.7 \times 4.6 \times 10^4}{1.9 \times 3.5 \times 10^3}$

$$= 5.79 \times 10 = 57.9$$

By using the cursor line to transfer from one pair of scales to another, it is possible to evaluate an expression involving a square or square root without having to write down intermediate values.

Example   Evaluate $1.9\sqrt{23.4}$.
i)   Set the cursor line over 23.4 on the A scale. (This gives $\sqrt{23.4}$ under the cursor line on the D scale.)
ii)   Move the slide to bring 1 on the C scale under the cursor line.
iii)   Move the cursor till the cursor line is over 1.9 on the C scale.
iv)   Read the answer on the D scale under the cursor line (9.19).

Using B, C and CI scales we can evaluate an expression such as $1/\sqrt{0.35}$ simply by aligning the cursor line over 35 on the B scale. Under the cursor line we now have $\sqrt{35}$ on the C scale and its reciprocal on the CI. Reading this as 169, we place the decimal point by inspection to get 1.69 as the answer.

Exercise B3
Each of the following expressions should be evaluated by slide rule. The

answers should then be checked by using tables or a calculator. (There are no answers for this exercise at the end of the book.)

1. 3.4×2.7

2. 3.45×2.75

3. $(4.35)^2$

4. $\sqrt{78.8}$

5. $1/3.25$

6. $7\sqrt{6.58}$

7. $(0.92)^2$

8. $\sqrt{673}$

9. $\dfrac{4.37 \times 2.69}{3.14}$

10. $\dfrac{10.7 \times 6.93}{(4.82)^2}$

11. $\dfrac{(2.59)^2}{\sqrt{3.05}}$

12. $\dfrac{4.37 \times 9.56 \times 7.83}{6.72 \times 5.19}$

13. $\dfrac{10.7 \times 0.98 \times 2090}{4.22 \times 11.3}$

14. $\dfrac{72.8 \times 20.02 \times 0.08}{3.78 \times 1.95}$

8 Using a calculator

8.1 Basic operations

With the older type of calculating machines it was necessary to move a lever into position for each digit of a number, then rotate the handle to enter the number. Rotation of the handle in one direction gave addition, in the reverse direction it gave subtraction. In such machines, multiplication was performed by repeated addition and division by repeated subtraction.

With modern electronic calculators, all arithmetical calculations become simple press-button operations. Numbers are entered by pressing the appropriate button for each digit in the number. By international agreement, the digits from 1 to 9 are arranged in a 3 x 3 block with $\boxed{7}$ $\boxed{8}$ $\boxed{9}$ on the top row, $\boxed{4}$ $\boxed{5}$ $\boxed{6}$ on the centre row and $\boxed{1}$ $\boxed{2}$ $\boxed{3}$ on the lowest row, with the zero $\boxed{0}$ either to the left or below. Simple calculators may have little more on the remaining buttons than the four basic operations of $\boxed{\times}$ $\boxed{\div}$ $\boxed{+}$ $\boxed{-}$ together with an equals $\boxed{=}$, a cancel or clear $\boxed{C}$ and either a decimal point $\boxed{.}$ or an exponent $\boxed{E}$.

The procedure varies according to the type of calculator you are using and you should refer to the instruction book to find the sequence appropriate to your particular machine.

Warnings

1. Before starting any calculation, clear the machine (including any memory).
2. Limit your answers to a reasonable number of significant figures (see section 5.1).
3. If using a battery model, switch off your machine between calculations (to conserve battery life).
4. If the display begins to fade, recharge or replace the batteries or the calculations may be affected.

8.2 Integral powers

These may be done as repeated multiplication of the base by itself, e.g. $(4.56)^3$ = $4.56 \times 4.56 \times 4.56 = 94.8$. Some calculators have a button for squares $\boxed{x^2}$

which can be used for squares and for even powers,

e.g. $(4.56)^4 = [(4.56)^2]^2 = 432$

$(4.56)^6 = [(4.56)^2 \times 4.56]^2 = 8.99 \times 10^3$

8.3 Reciprocals
Some calculators have a button for reciprocals $\boxed{1/x}$ which involves merely entering the number then pressing this function button. If your calculator does not have this facility, then you may obtain the reciprocal of a number by dividing the number into 1.

8.4 Evaluations
Your calculator will help you both in the arithmetical calculations arising from your work in other technical subjects and also in evaluation from formulae (see section 12.1). You are advised to practise using your calculator until you can operate it quickly and accurately. Try using it to check results obtained by using your slide rule or four-figure tables.

8.5 Advantages and disadvantages
The electronic calculator possesses obvious advantages of easy and accurate operation. Its dependence upon some source of electrical power may prove a disadvantage in certain situations, so the use of slide rules and four-figure tables should not be abandoned. A set of tables is probably the cheapest of these calculation aids, but many slide-rule calculations can be carried out on the spot in a workshop without the need for paper and pencil.

Exercise B4
Use a calculator to evaluate each of the following expressions:
1. $3972 + 4165 + 2971 + 385$
2. $2.763 + 0.512 + 11.37 + 7.961$
3. $4734 + 5296 - 3488$
4. $10.356 - 1.983 - 2.497$
5. 283×471
6. $2.735 \times 0.625 \times 1.024$
7. $74.35 \div 11.59$
8. $(6.594)^2$
9. $1/2.151$
10. $47.1 \times (0.763)^2$
11. 5% of £7360
12. 2% of \$375
13. $\dfrac{21.73 \times 4.812}{37.16}$
14. $(2.594)^2 - (1.762)^2$
15. $(0.968)^3$
16. $(1.6754)^4$
17. $\dfrac{41.78}{2.421} - \dfrac{0.736}{0.566}$
18. $1/(0.877)^2$

C Algebra

9 Basic notation and rules of algebra

9.1 Use of letters

In mathematics, letters are used for several purposes.

i)   To represent units, e.g. m for metres, N for newtons, etc., and in such cases an abbreviated form of the unit is indicated by the letter or letters used.

ii) To represent known constants, e.g. we use g for $9.806\,65$ m/s^2 and π for $3.141\,592\,65\ldots$

iii) To represent unknown constants, e.g. a straight line at $45°$ to the x-axis on a graph would have the equation $y = x + c$, where c is constant.

iv) To represent quantities in a formula, e.g. $A = l \times b$, where A is area, l is length and b is breadth.

v)   To represent variables, e.g. $\sin^2 x + \cos^2 x = 1$, where x can be any angle (in degrees or radians).

To avoid ambiguity, care should be taken not to represent an unknown quantity by a letter which could be mistaken for the symbol for a particular constant or unit.

It is common practice to use letters from the first part of our alphabet to represent constants and letters from the second half of the alphabet for variables. For example, in the equation $ax + by + c = 0$, letters a, b, c represent constants, x and y are variables.

Coefficients

In the expression $2x + 3y - 4$, the number, 2, which multiplies x, is called the *coefficient* of x. Similarly, the coefficient of y is 3. The remaining number is -4 and this could be described as the *numerical* term, to distinguish it from the other two parts which each contain a variable.

In the quadratic equation $6x^2 - 5x + 1 = 0$, the coefficient of x^2 is 6, the coefficient of x is -5, and the numerical term is $+1$. In some cases a coefficient may be unknown and be represented by a letter.

In the general expression for a quadratic equation we write $ax + bx + c = 0$, where $a, b,$ and c are all constants. In this case the coefficient of x^2 is a.

Coefficients can contain both letters and numbers, so that if we had a term $6ax^3$ we would say that the coefficient of x^3 was $6a$.

Like and unlike terms

Terms such as $4x$ and $3x$ have the same variable but different coefficients. They are known as *like terms* and can be added together (or subtracted) so that $4x + 3x = 7x$ and $4x - 3x = x$.

If we had $4x$ and $3y$, they would be *unlike* because they have different variables. Similarly, $2x^2$ and $3x$ are unlike terms. This means that neither $4x + 3y$ nor $2x^2 + 3x$ can be reduced to a single term.

When letters are used to represent units, a similar situation exists, for we can add 3 grams to 4 grams to give $3\,g + 4\,g = 7\,g$, but we cannot add quantities expressed in different units. Sometimes we can bring the quantities to the same units and then combine them, so that $2\frac{1}{2}$ kg plus 800 g would be 2500 g + 800 g = 3300 g or 2.5 kg + 0.8 kg = 3.3 kg.

9.2 Simplifying expressions

a) We can apply the four operations of addition, subtraction, multiplication and division. For example,

$$4a + 2a = 6a \qquad 3x^2 - x^2 = 2x^2$$

$$5p \times 2q = 10pq \qquad 6y \div 3 = 2y$$

b) We can apply the commutative, associative and distributive laws (see sections 1.2 to 1.4):

$$a + b = b + a$$

$$ab = ba$$

$$a + (b + c) = (a + b) + c$$

$$a(bc) = (ab)c$$

$$a(b + c) = ab + ac$$

$$\frac{a + b}{c} = \frac{a}{c} + \frac{b}{c}$$

c) We can apply the laws of indices (see section 3.2):

$$a^m a^n = a^{m+n}$$

$$\frac{a^m}{a^n} = a^{m-n}$$

$$(a^m)^n = a^{mn}$$

$$a^{-n} = 1/a^n$$

d) We shall observe the precedence rules (see section 1.11):

Brackets
Of
Divide
Multiply
Add
Subtract

Exercise C1

1. Of the following 10 expressions, (a) pick out the two which cannot be simplified further, (b) simplify the others as far as possible.

$3b + 2b$ $5x - 3x + x$ $4x^2 - x$ $2x \times 5x$ $4 \times 3y \times y^2$

$7z^2 - z^2$ $4a + 5b - 7c$ $\frac{1}{3}$ of $9x^3$ $6t \times 2t^2 - t^3$ $\frac{1}{3}v^2 + \frac{1}{6}v^2$

2. Use the laws of indices to simplify the following expressions:

$a^4 \times a^2$ $2b^2 \times b^3$ c^5/c^2 $6d^7/2d^3$

$(x^3)^4$ $(2y^2)^3$ 4^{-1} $(3z)^{-2}$

3. Put each of the following expressions into its simplest form:

$2 + a + 3a - 1$ $5b - 4 + 2b - 6$ $1.6 + 0.4x - 2.8 + 1.9x$

$(2y + 5) - (y + 3)$ $2(z^2 - 3) + 3(z^2 + 2)$

4. Subtract $4a - b$ from $7a + 2b - c$.
5. Simplify to a single fraction

$$\frac{3}{x} + \frac{x + 5}{2x} - \frac{1}{2}$$

6. Simplify $\frac{1}{2}(3t^2 - t + 2) - \frac{1}{4}(2t^2 - 2t + 3)$.

10 Factorizing algebraic expressions

10.1 Multiplication of brackets
The distributive law for multiplication gives us

$$a(b + c) = ab + ac$$

and we can substitute any value (numerical or otherwise) for the letters a, b, c.

Examples $2(3 + x) = 6 + 2x$

$$p(q - 2) = pq - 2p$$

$$4x(x + 5) = 4x^2 + 20x$$

When multiplying one bracket by another, the terms in the second bracket must each be multiplied by each term in the first bracket. It may then be possible to simplify the subsequent expression by collecting like terms.

Examples
$$(x + 1)(x + 2) = x^2 + 2x + x + 2 = x^2 + 3x + 2$$

$$(y - 2)(3y - 4) = 3y^2 - 4y - 6y + 8 = 3y^2 - 10y + 8$$

$$(a + b)^2 = (a + b)(a + b) = a^2 + ab + ab + b^2 = a^2 + 2ab + b^2$$

$$(a - b)^2 = (a - b)(a - b) = a^2 - ab - ab + b^2 = a^2 - 2ab + b^2$$

$$(a + b)(a - b) = a^2 - ab + ab - b^2 = a^2 - b^2$$

The last three examples are particularly important and the final one is known as 'the difference of two squares'. The standard form $a^2 - b^2 = (a + b)(a - b)$ should be memorized.

10.2 Factorization

If two or more terms have a common element, this can be extracted as a factor. The process is indicated by the reverse of the distributive law for multiplication, since if $a(b + c) = ab + ac$, then $ab + ac = a(b + c)$.

Examples
$$2x + 2y = 2(x + y)$$
$$2x - 4y = 2(x - 2y)$$
$$x^2 - 4x = x(x - 4)$$
$$2a + ab - 3ac = a(2 + b - 3c)$$
$$2x^3 + 4x^2 - 6x = 2x(x^2 + 2x - 3)$$

We see that the factor taken out of the bracket can be a number, a letter, or both together. It is also possible to take out a factor in a bracket.

Examples
$$a(x + 2) + b(x + 2) = (x + 2)(a + b)$$
$$x(a - 3) + y(a - 3) = (a - 3)(x + y)$$
$$q(2 - p) - r(2 - p) = (2 - p)(q - r)$$
$$4(x + y) - 2x(x + y) = 2(x + y)(2 - x)$$
$$3(a + b) + (x + 1)(a + b) = (a + b)(3 + x + 1) = (a + b)(4 + x)$$

In some cases, terms have to be grouped first before the presence of a common factor is apparent.

Examples
$$ax - ay + bx - by = a(x - y) + b(x - y) = (x - y)(a + b)$$
$$3x - 3xy + y - y^2 = 3x(1 - y) + y(1 - y) = (1 - y)(3x + y)$$

Note that each of these two examples could have been factorized in a different order without affecting the result.
$$ax - ay + bx - by = x(a + b) - y(a + b) = (a + b)(x - y)$$
$$3x - 3xy + y - y^2 = 3x + y - y(3x + y) = (3x + y)(1 - y)$$

This demonstrates that $(x - y)(a + b) = (a + b)(x - y)$, which shows that the commutative law for multiplication also applies to brackets.

Exercise C2

1. Simplify $2 + x + 3 + 5x - 4 - 2x$.

2. Simplify $\dfrac{2x - 3y + 1}{6} + \dfrac{y}{2}$.

3. Add $x - 2y + 3$ to $3x + 2y - 1$.
4. Subtract $a - 2b$ from $4a + b$.
5. Simplify $(a + b)^2 + (a - b)^2$.
6. Divide $6x^2y$ by $2xy$.
7. Factorize $9x - 6x^2$.
8. Factorize $5(p - q) - x(p - q)$.
9. Factorize $a^2 + ab + ac + bc$.
10. Factorize $2ab - b + 6a - 3$.
11. Factorize $7x - xy + 14 - 2y$.
12. Factorize $ax + ay - a + bx + by - b$.

11 Algebraic solution of equations

11.1 Expression or equation?
In the last exercise, none of the questions contained an 'equals' sign. In that exercise we were concerned with simplifying or factorizing *expressions*. Note that we were not allowed to make an expression simpler by removing any part of it; we could only rearrange its terms. The expression at the end had to be equivalent to the expression we had started with, even though its form had changed.

An *equation* represents an exact balance between what is on one side of the 'equals' sign and what is on the other side. It is essential to maintain this balance while working on an equation, so if we add a certain quantity to one side of the equation we have to add an equal quantity to the other side of the equation. If we double one side, we must double the other, etc.

11.2 Operating on an equation
It is quite obvious that if $x + 3 = 5$ then $x = 2$, but how do we arrive at this answer? Logically we should proceed as follows:

write down the original equation	$x + 3 = 5$
subtract 3 from each side	$x + 3 - 3 = 5 - 3$
which simplifies to	$x = 2$

Now the same result could have been obtained simply by transferring $+3$ to the other side of the equation and changing its sign to -3, but if we choose to do this we should realize that it is only a shortened form of the more detailed process shown above.

For an equation such as

$$\tfrac{2}{3}x = 8$$

we can take the following steps:

multiply each side by 3 $\qquad 2x = 24$

divide each side by 2 $\qquad x = 12$

although with practice we could do these two steps in one by multiplying each side by $\frac{3}{2}$.

We have seen that we have to maintain the balance of the equation while performing the operations of addition, subtraction, multiplication or division. The same principle of maintaining the equality applies to any other operations such as squaring both sides, taking the square root or the reciprocal of each side, or even operations beyond the present level (such as taking the logarithm of each side).

11.3 Solving simple equations

A simple linear equation has only one unknown and this is not raised to any power. Solving the equation means finding the particular numerical value of the unknown which makes the equation balance exactly. This usually requires the equation to be manipulated until we have the unknown left by itself on one side and the appropriate number on the other. This number is then called the solution.

Example Solve $3z - 4 = 2z + 1$.

Add 4 to each side (to bring all the numerical terms to the right):
$$3z = 2z + 1 + 4$$
Subtract $2z$ from each side (to bring all the z terms to the left):
$$3z - 2z = 1 + 4$$
$$z = 5$$

In this case we have found that $z = 5$ is the solution of the equation $3z - 4 = 2z + 1$. We say that this value 'satisfies' the equation. You can check this by putting z equal to 5 in each side of the original equation and showing that this gives 11 in each case.

It is possible for a linear equation to contain a bracket, but this can usually be removed by multiplying out.

Example Solve $5(2x - 1) - 6 = 9$.

Add 6 to each side	$5(2x - 1) = 15$
Multiply out	$10x - 5 = 15$
Add 5 to each side	$10x = 20$
Divide by 10	$x = 2$

We check this by putting $x = 2$ in $5(2x - 1) - 6$ which gives us $5 \times 3 - 6 = 9$, the same as the right side, so $x = 2$ is the required solution.

A linear equation could also involve fractions, and in such cases it is usually best to clear the fractions first by multiplying through by the LCM of the denominators.

Example   Solve $\dfrac{2x+1}{4} = x - \dfrac{2x-1}{3}$.

Multiply through by 12, remembering that the horizontal line in the two fractions is equivalent to a bracket:

$3(2x + 1) = 12x - 4(2x - 1)$

Remove the brackets:

$6x + 3 = 12x - 8x + 4$

Simplify:

$6x + 3 = 4x + 4$

Subtract $4x$ from each side:

$2x + 3 = 4$

Subtract 3 from each side:

$2x = 1$

Divide by 2:

$x = 0.5$

11.4 Equations from practical work

In an experiment to measure the expansion of a metre bar of metal, the formula $L = L_0(1 + \alpha t)$ gave the equation $1004 = 1000 (1 + 20\alpha)$ after substituting all the measurements. The solution of this equation gives the value of α (the coefficient of linear expansion of the bar).

Write the equation as	$1000(1 + 20\alpha) = 1004$
Multiply out	$1000 + 20\,000\alpha = 1004$
Subtract 1000	$20\,000\alpha = 4$
Divide by 20 000	$\alpha = \dfrac{4}{20\,000}$
	$\alpha = 2 \times 10^{-4}$

Sometimes there is no given formula and an equation has to be constructed from the basic information.

Example   How many kilograms of an ingredient costing \$9 per kg should be added to 10 kg of a second ingredient costing \$3 per kg to give a mixture worth \$5 per kg?

Let x be the number of kg of the expensive ingredient. Added to 10 kg of the cheaper ingredient this gives $(10 + x)$ kg of mixture.

The cost equation is therefore

$9x + 3 \times 10 = 5(10 + x)$

$9x + 30 = 50 + 5x$

$4x = 20 \qquad \therefore x = 5$

The mixture therefore requires 5 kg of the expensive ingredient.

Exercise C3
1. Solve each of the following equations:

$2x - 5 = x + 2$

$7y + 1 = 9 - y$

$3 + z = 1 - z$

2. By substituting the value $x = 2$, check whether this value satisfies the equation $2(x - 1) + 3 = 3(4 - x) - 1$.
3. Solve the following equations:

$5p - 3 = 12 - 2p$

$3(q - 2) = 21$

$5(6 - r) = 20$

$4(5 - 2s) = 12$

$2(7 + 3t) = 8t$

4. Solve the equation $\frac{1}{3}(2z + 1) = \frac{1}{2}(z + 2)$.
5. Find the value of x which satisfies the equation

$2.5(x + 2) = 3(4.6 - x)$

6. Solve the equation $\frac{a + 2}{4} + \frac{2a - 3}{6} = \frac{a + 3}{3}$.
7. Solve $5(x - 1) - 4(x - 2) = 2(4x - 9)$.
8. Solve $\frac{1}{2}(x + 2) - \frac{1}{3}(x + 1) = \frac{1}{4}(x + 6) - 1$.

11.5 Simultaneous equations
A single equation is sufficient to find the value of one unknown, as we have seen. Two equations are required for two unknowns, three for three unknowns, etc., these being reduced to an equation in one unknown by elimination or substitution. These equations, which have to be solved together to find the values which fit both equations, are known as simultaneous equations.

Elimination
Consider the two simultaneous equations:

$3a - 2b = 5$ . . . . (i)

$a + 2b = 7$ . . . . (ii)

Adding, $4a + 0 = 12$

b has thus been eliminated, leaving an equation in a alone.

$\therefore \quad a = \frac{12}{4} = 3$

Alternatively, multiply equation (ii) by 3 to make the coefficient of *a* the same in each equation:

$$3a + 6b = 21$$

equation (i) $\quad 3a - 2b = 5$

Subtracting, $\quad 0 + 8b = 16$

which is an equation involving only *b*.

$$\therefore \quad b = \frac{16}{8} = 2$$

Substitution

Taking the same equations:

$$3a - 2b = 5 \qquad . \quad . \quad . \quad . \quad . \quad . \quad . \quad . \quad . \quad \text{(i)}$$
$$a + 2b = 7 \qquad . \quad . \quad . \quad . \quad . \quad . \quad . \quad . \quad . \quad \text{(ii)}$$

From equation (ii), $a = 7 - 2b$ and this expression for *a* can be substituted in equation (i) to give

$$3(7 - 2b) - 2b = 5$$

Multiplying out,

$$21 - 6b - 2b = 5$$
$$8b = 16$$
$$\therefore \quad b = 2$$

Similarly, from equation (i), $2b = 3a - 5$ which can be substituted in equation (ii) to give

$$a + (3a - 5) = 7$$

$$4a - 5 = 7$$

$$a = 3$$

Although it is possible to solve such simultaneous equations entirely by elimination or entirely by substitution, it is usually quicker to use a combination of the two methods.

Example Two quantities *x* and *y* are connected by the relationship $y = mx + c$, where *m* and *c* are constants. Given that $y = 4$ when $x = 3$ and that $y = 2$ when $x = -2$, find the values of *m* and *c*.

Putting in the values gives the two equations:

$$4 = 3m + c \qquad . \quad . \quad . \quad . \quad . \quad . \quad . \quad . \quad . \quad \text{(i)}$$

$$2 = -2m + c \qquad . \quad . \quad . \quad . \quad . \quad . \quad . \quad . \quad . \quad \text{(ii)}$$

Subtracting to eliminate *c*,

$$2 = 5m$$

$$m = 0.4$$

Substituting this value in equation (i),

$4 = 1.2 + c$

$c = 2.8$

After substituting in one equation it is always a good check to ensure that the values obtained do satisfy the other equation; e.g. for equation (ii)

right-hand side $= c - 2m = 2.8 - 0.8 = 2 = $ left-hand side

The methods of elimination and substitution may be applied to any number of simultaneous linear equations.

Exercise C4

1. Solve the simultaneous equations $2a + 3b = 7$, $a + b = 3$.
2. Find the values of x and y which satisfy the following equations:

$x + 3y = 2$ and $2x - y = 11$

3. If $4s - t = 5$ and $s + 4 = t$, find the values of s and t.
4. Solve for p and q:

$p - q = 1$ $\qquad 2p - 3q = 5$

5. Solve the pair of simultaneous equations

$2.4x - 3y = 6$ $\qquad x + 1.5y = 8$

6. Find m and n if $4m + 5n = 10$ and $m + 2n = 3.7$.
7. Given the relationship $v = u + at$, put in the values of $v = 50$ and $t = 10$ to form an equation involving a and u. Put in the values $v = 80$ and $t = 20$ to form a second equation.

   Solve the two as simultaneous equations and so find the values of the constants a and u.
8. If $V = E + IR$ and experimental results show that $V = 10$ when $I = 2$ and $V = 13$ when $I = 3$, substitute these values and solve the resulting simultaneous equations to find the values of the constants E and R.

12 Use and transposition of formulae

12.1 Evaluation from formulae

In addition to the formulae for volumes and surface areas in section 18, there are many other formulae in mathematics and physical science which we have to be able to use. Some are straightforward products such as

$V = I \times R$ $\qquad$ voltage = current x resistance (Ohm's law)

$W = F \times D$ $\qquad$ work = force x distance

$C = \pi \times D$ $\qquad$ circumference = π x diameter

$V = f \times \lambda$ $\qquad$ velocity = frequency x wavelength

49

Example Find the potential difference (voltage) across a 3 ohm resistance carrying a current of 0.2 amperes.

$$V = I \times R = 0.2 \times 3 = 0.6 \text{ volts}$$

Other formulae may involve reciprocals, brackets, squares or square roots, etc., but once the known values are substituted in the appropriate formula it becomes a simple equation with only one unknown and we use the methods of the previous section to solve the equation.

Example Given that the formula for the focal length of a convex lens is

$$\frac{1}{f} = \frac{1}{u} + \frac{1}{v}$$

find f when $u = 50$ mm and $v = 200$ mm.

$$\frac{1}{f} = \frac{1}{50} + \frac{1}{200} = \frac{4+1}{200} = \frac{5}{200} = \frac{1}{40}$$

$$\therefore \quad f = 40 \text{ mm}$$

12.2 Transposition of formulae

A formula, such as $V = \pi r^2 h$ for the volume of a cylinder, expresses one quantity (V) in terms of other quantities (r and h) and sometimes a constant (π). It is conventional that the quantity we are trying to find (V) should be on the left and everything else on the right. Dividing each side of the formula by πr^2 gives

$$h = \frac{V}{\pi r^2}$$

and this is a formula giving the height of a cylinder in terms of the volume and radius. Having changed the original expression of the formula into this new form, we may call this process a *transformation* of the formula or, since we have changed the position of the various symbols, we may call it a *transposition* of the formula.

The third possible form of this particular formula would give the radius in terms of the volume and height.

From $V = r^2 h$

dividing by h gives $r^2 = \dfrac{V}{\pi h}$

Taking the square root, $r = \sqrt{\dfrac{V}{\pi h}}$

In section 11.4, we used the formula $L = L_0(1 + \alpha t)$ to find the value of α in a particular example. In that instance we simply substituted the values for L, L_0 and t and then worked out the value of α. We could have transposed the formula first and then substituted the values.

$$L_0(1 + \alpha t) = L$$

Dividing by L_0

$$1 + \alpha t = \frac{L}{L_0}$$

Subtracting 1,

$$\alpha t = \frac{L}{L_0} - 1$$

Dividing by t,

$$\alpha = \frac{1}{t}\left(\frac{L}{L_0} - 1\right)$$

Into this new formula for α we now substitute $L = 1004$, $L_0 = 1000$, $t = 20$, to give

$$\alpha = \frac{1}{20}\left(\frac{1004}{1000} - 1\right)$$

$$\alpha = \frac{1}{20}(1.004 - 1)$$

$$\alpha = \frac{0.004}{20} = \frac{4}{20\,000}$$

$$\alpha = 2 \times 10^{-4}$$

In general, for a *single* set of values it is better to use the formula in its given form and then work out the value of the remaining quantity after substitution, but if we have a number of different sets of values all to be substituted, then it is better to transpose the given formula into its most convenient form for our repeated use. This applies particularly to repeated calculations with computers and calculating machines.

Exercise C5

1. Given $v = u + at$, find v when $u = 50$, $a = 9.8$, $t = 30$.
2. If $J = mc(t_2 - t_1)$, find J when $m = 10$, $c = 0.39$, $t_2 = 323$, $t_1 = 273$.
3. Given that $\frac{1}{f} = \frac{1}{u} + \frac{1}{v}$, find f when $u = 150$ and $v = 300$.
4. If $\frac{1}{R} = \frac{1}{r_1} + \frac{1}{r_2}$, find R when $r_1 = 0.5$ and $r_2 = 1.5$.
5. From the formula $W = F \times D$, find F when $W = 1000$ and $D = 50$.
6. What is the value of R if $V = 240$ when $I = 12$ in the formula $V = IR$?
7. If $v = f \times \lambda$, what is λ in terms of v and f?
8. From the formula $v = u + at$, find a formula for a in terms of t, u and v.
9. Given $v^2 = u^2 + 2ax$, derive a formula for x in terms of the other quantities.
10. If $V = \frac{4}{3}\pi r^3$ and $r = \frac{1}{2}D$, find a formula for V in terms of D.

D Graphs

13 Graphs illustrating one-to-one relationships

13.1 Conversion of data

The adoption of SI units has considerably reduced the problems associated with different systems of units, but it is still necessary to convert some data from one set of units to another. When we go to another country on holiday we have to convert currency, and it would be so much easier if we could use conversion rates like £1 = $2 = 10F! However, conversions are rarely so simple.

For many years there have been several scales for measuring temperature, but the Fahrenheit scale is gradually becoming obsolete. Wherever possible we use the Kelvin scale, but temperatures are frequently measured in degrees Celsius (Centigrade) because this is more convenient for everyday use. Since $0°C = 273$ K $= 32°F$ and $100°C = 373$ K $= 212°F$, we derive the relationships

$$t°C = (t + 273) \text{ K} \quad \text{and} \quad t°C = (1.8t + 32) °F$$

Such relationships can be used to convert a temperature on one scale to the corresponding temperature on one of the other scales. For example, if a person's temperature was measured as 37.0°C, then this would be 310 K or 98.6°F.

13.2 Parallel scales

A greenhouse thermometer may have parallel scales as shown in fig. D1 so that the temperature may be read either in degrees Fahrenheit or in degrees Celsius as required.

A speedometer in a car may have parallel scales as shown in fig. D2(a) or, if a circular dial fits in better with the dashboard layout, the two scales could be placed in the form of a circular arc as shown in fig. D2(b), but in either case the speed can be read both in miles per hour and in kilometres per hour.

Similar arrangements of parallel scales can be used for conversion of gallons to litres or, for tyre pressures, between pounds per square inch and newtons per square metre.

Such parallel scales are most useful when two sets of values are directly proportional one to the other. In mathematics we write this as $y \propto x$, but it is usually convenient to write this as an equation such as $y = kx$, where k is constant. (See section 13.5 for examples.)

13.3 Straight-line graphs

For the simplest kind of graphs, we draw two axes at right angles intersecting at the origin O. The horizontal axis OX is often referred to as the x-axis, and the vertical axis OY as the y-axis. The position of any point on the graph is

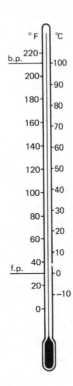

Fig. D1 Temperature scales

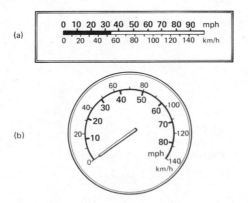

Fig. D2 Speedometer scales

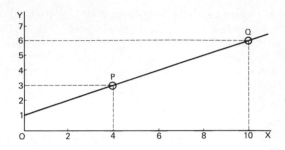

Fig. D3

then fixed if its distance from each axis is given, such distances being known as *co-ordinates*.

In fig. D3, the point P on the straight-line graph corresponds to the value 4 on the x-axis and to the value 3 on the y-axis. We say, therefore, that $y = 3$ when $x = 4$. Similarly, from point Q we see that $y = 6$ when $x = 10$.

The scales we mark on our axes do not necessarily always start from zero. With the Kelvin scale for temperature, for example, a range of 250 K to 400 K will cover all the experiments we are likely to do in the laboratory. If we were using a graph of body temperature, we might well need only a range of values from 308 K to 318 K, so we could mark values such as 308, 310, 312, 314, 316 and 318 K at equal intervals along the appropriate axis and our graph would start from 308 K and not from zero.

Sometimes negative values are necessary and fig. D4 shows a temperature-conversion graph for converting readings in degrees Celsius to values in degrees Fahrenheit.

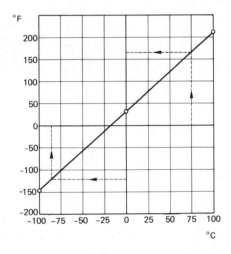

Fig. D4

54

Intermediate values may now be determined from the graph without using the formula. Thus, to convert 75°C to degrees Fahrenheit, we simply trace the 75 °C line up until it crosses the graph, then read across horizontally on to the degrees Fahrenheit axis to get the value 167 °F. Similarly, to change −121 °F to degrees Celsius, trace the level −121 °F across until the graph is reached, then follow up vertically to the value −85 °C.

13.4 Mappings

The students in a certain class make a list of the places where each of them was born. This set of place names and the set of students' names on the class register are obviously related. For any student we can find an appropriate place name. Finding which element of the second set corresponds to a particular element of the first set is called *mapping*. It is possible that every student was born in a place different from every other student, and in that case both sets would contain the same number of elements. Matching each place to the appropriate person would then be an example of one-to-one mapping. If several students were born in the same town the relationship between those students and their birthplace would be an example of a many−one mapping.

When two variables are connected by a mathematical relationship such as an equation or a formula, we say that one variable is a *function* of the other. The examples we have already considered, such as currency conversion or changing temperature units etc., are examples of mappings which are functions. In each case a value on one scale is mapped on to the corresponding value on the other scale.

13.5 Proportional quantities

The formula $C = \pi D$ tells us that the circumference of a circle is a function of its diameter (see section 17.2). In this formula, D is a variable and can be given any value from microscopic (such as the thickness of a human hair) to astronomic (such as the diameter of the sun). D is thus the *independent variable* and, since C is a function of D, the value of C depends on the value of D, so C is the *dependent variable*. The circumference is proportional to the diameter ($C \propto D$) which means that there is a direct relationship between C and D such that

circumference = constant x diameter

and in this case the constant is the value we refer to as π (approximately 3.1416).

Some of the laws in physical science are concerned with the relationship between quantities which are directly proportional. For example:
a) Hooke's law for the stretching of wires, springs, etc. gives the relationship

extension $\propto$ load

b) Ohm's law implies that if we are concerned with the current flowing through a fixed resistance under a potential difference (voltage) then

current $\propto$ voltage

c) Charles' law tells us that at constant pressure the volume of a fixed mass of gas is directly proportional to the absolute temperature (i.e. the temperature on the Kelvin scale).

It should not be assumed that *all* laws in physical science involve direct proportionality. Boyle's law, for example, tells us that at constant temperature the volume of a fixed mass of gas is *inversely* proportional to the pressure upon it,

i.e.   volume $\propto \dfrac{1}{\text{pressure}}$

When one quantity is directly proportional to another, a simple straight-line graph can be drawn to illustrate the relationship.

Exercise D1
1. Working on the basis that one Canadian dollar is equivalent to 40 Belgian francs, construct parallel scales suitable for the conversion of any amount up to 1000 Belgian francs into Canadian dollars.
2. Draw a straight-line graph for the conversion of Japanese yen into Italian lire. Use the vertical axis for yen and the horizontal axis for lire and assume that 10 yen are equivalent to 32 lire. Use scales up to 50 000 yen (160 000 lire). Draw lines to illustrate how you would use your graph to convert 100 000 lire to yen and show clearly the value you obtain.
3. Given that 1 radian is equivalent to 57.3 degrees, construct a straight-line graph for converting angles to radians up to 360°. Use your graph to read off how many radians are equivalent to 200°.
4. With reference to section 13.5, explain why Hooke's law ceases to apply if a wire is overloaded. Comment also on how far Ohm's law can be expected to apply as the current flowing through a wire is gradually increased.

13.6 Axes and co-ordinates
As indicated in section 13.3, it is conventional to use a pair of axes at right angles to one another, the values for the independent variable being along the horizontal axis and for the dependent variable along the vertical axis. We usually use x for the independent variable and y for the dependent variable, so that the resulting graph expresses y in terms of x and indicates that y is a function of x.

Any point in the space between the axes can now be defined by its co-ordinates. This pair of values is usually put in brackets with a comma between, and these values are often referred to as (x, y) co-ordinates since the value of the independent variable, x, is put first. Thus, in fig. D3, point P has co-ordinates $(4, 3)$ and point Q has co-ordinates $(10, 6)$.

13.7 Choice of scales
A large graph should be more accurate than a small one, so it is essential to make full use of the graph paper. While it is convenient to start one or both scales from zero, it is better not to do so if it results in crowding the points into one small corner of the graph paper.

On each axis we have to cover a certain range within a limited space, but it is better to choose scales which fit the subdivisions on the graph paper, so we let one unit represent,1, 2, 5 or 10 rather than 3, 7 or 9.

13.8 Labelling
The graph should have a title to explain what it is all about. Each axis should also be labelled so that it is clear what the numbers represent. This implies not only stating the variable on each axis but also putting in any appropriate units. Sufficient figures should be marked along the axis at equal intervals to help in reading the graph, but overcrowding should be avoided.

13.9 Plotting
Given the co-ordinates, a particular point is easily located in terms of its distance along the horizontal and vertical axes. Having found precisely where the point should be, we mark the place in one of two ways. Avoiding the temptation to use the rough 'X marks the spot' technique, we either place a very fine dot on the exact location with a circle round it to make it easy to find, or we use fine vertical and horizontal lines (like a 'plus' sign) with the intersection exactly over the required location.

Either method can be refined to indicate the accuracy of an experimental observation. In the first case, the diameter of the circle drawn round a particular point indicates the possible error limits of measurement. In the second case the extent of possible inaccuracy can be indicated by the lengths of the short horizontal and vertical lines which form the cross and which intersect over the optimum value.

13.10 Drawing graphs
At this stage we are concerned with straight-line graphs, which are the easiest to draw since we can use a ruler or other straight edge. The line should be drawn in pencil, never with a ball-point or fibre-tip pen. Ideally it should pass through all the points which have been plotted, but we have to realise that it does not necessarily pass through the origin as well unless (0,0) is one of the points we have plotted. In experimental graphs in which we have used circles round the points to indicate possible error limits, we should find that our graph line will intersect all such circles if we have deduced the right relationship between the two variables — see fig. D5 which shows the relationship between the speed of an electric motor (N revolutions per second) and the applied voltage (V volts) in accordance with the following experimental results:

V	10	20	30	40	50	60
N	18	36	55	71	90	108

13.11 Reading values
Taking any point on a graph, traversing vertically and horizontally from the point on to the axes will give us the co-ordinates of the point.

In section 13.3 and fig. D4 we saw how a particular value on one axis is connected with the corresponding value on the other axis via a point on the

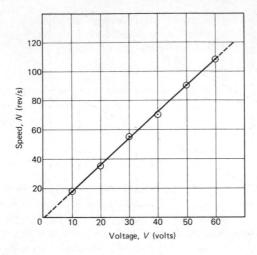

Fig. D5 Electric-motor experiment

graph. With the same method, we can use the graph of fig. D5 to estimate what speed of rotation will result from a certain voltage, or what voltage will produce a particular speed. If, for example, we wanted to know what voltage to apply to get a speed of rotation of 45 revs per second, we simply trace across horizontally at the value of 45 on the N-axis until we reach the line of the graph, then drop vertically on to the V-axis to find that the required value is exactly 25 volts.

13.12 Gradient

In fig. D6, the lower straight-line graph has the equation $y = 0.5x$ (which could also be written as $y = \frac{1}{2}x$, or $2y = x$, or $x = 2y$). The virtue of writing the equation in the form $y = 0.5x$ is that the coefficient of x (in this case 0.5) gives us the gradient or slope of the graph.

Consider the section of the graph from O (the origin) to P (the point with co-ordinates $x = 4$, $y = 2$). By the time it reaches P, the graph has risen 2 units vertically while moving 4 units horizontally. Its slope is thus $\frac{2}{4} = 0.5$. If we consider the section from O to Q, we find the gradient to be $\frac{3}{6} = 0.5$. The areas between OP and the x-axis and OQ and the x-axis are similar triangles (see section 16.6). We conclude that the gradient of a straight-line graph is constant.

For the second straight-line graph in fig. D6 the equation is $y = 0.5x + 3$. Here also the coefficient of x is 0.5, so this graph has the same slope as the lower graph $y = 0.5x$. The two graphs are parallel, having the same gradient.

From the equation $y = 0.5x + 3$, we can see that when $x = 0$, $y = 3$. The graph therefore cuts the y-axis at $(0,3)$ and the +3 in its equation gives us the distance from the origin at which the line cuts the y-axis. This length is known as the *intercept*.

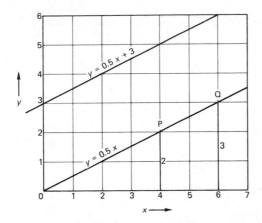

Fig. D6 Graphs with positive gradient

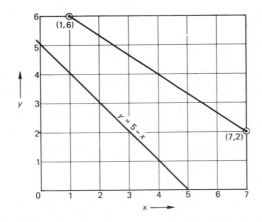

Fig. D7 Graphs with negative gradient

The graphs in fig. D6 have positive gradient, sloping upwards from left to right. The graphs in fig. D7 have negative gradient.

In fig. D7, the lower graph has a gradient of -1 and the intercept on the y-axis is $+5$, so its equation is $y = 5 - x$.

In general, any straight-line graph will have an equation of the form $y = mx + c$, where m is the gradient and c is the intercept. We can find the equation of the second graph in fig. D7 by substituting the known co-ordinates (x and y values) in the equation $y = mx + c$ and solving the resulting pair of simultaneous equations to find the appropriate values of m and c.

Since $y = 6$ when $x = 1$, $6 = m + c$

Since $y = 2$ when $x = 7$, $2 = 7m + c$

By subtraction,	$4 = -6m$	$m = -\frac{2}{3}$
By substitution,	$6 = -\frac{2}{3} + c$	$c = 6\frac{2}{3}$
So the equation is	$y = -\frac{2}{3}x + 6\frac{2}{3}$	
which may be written as	$2x + 3y = 20$	

Exercise D2

1. The following figures were obtained from an experiment to verify Ohm's law:

I (in amperes)	1.2	2.0	2.6	3.2	4.0
V (in volts)	6	10	13	16	20

   Plot a graph with values of V on the vertical axis and I on the horizontal axis. Show that the points lie on a straight-line graph. Find the gradient of the graph and state what this represents in Ohm's law.

2. Plot a graph of effort (E) against load (W) for the following set of results. (The forces shown are measured in newtons.)

E	8	11.0	16	20.2	24	28
W	10	20	30	40	50	60

   Find the gradient of the graph. Add lines to your graph to show how you would find the value of W when E is 25 newtons.

3. Find the equation of the straight line which joins the origin to the point (3,6).

4. From the following table of values, verify graphically that $R = KL$, where R is the resistance in ohms of a wire of constant diameter of length L in millimetres.

R	2	3	4	5	7
L	100	150	200	250	350

   Find the value of the constant K from the gradient of your graph.

5. Given the relationship $V = 9.8t$, where t is in seconds and V in metres per second, find the values of V for t values of 0, 2, 4, 6, 8 seconds and plot a graph of V against t.

6. In an experiment, readings are taken of temperature $T\,°C$ after time S seconds. It is known that T and S are connected by a relationship $T = a + bS$. Given that the temperature is $10°C$ after 5 seconds and $20°C$ after 15 seconds, find the values of the constants a and b.

7. A straight-line graph is given by the equation $y = mx + c$. Given that $y = 4$ when $x = 2$, and that $y = 7$ when $x = 8$, find the values of the constants m and c. Hence find the value of x which makes $y = 5$.

E   Statistics

14 Collecting, presenting and interpreting statistical data

14.1 Collecting data

We collect statistical data by counting or by measuring.

Figures gathered by counting individual items are known as *discrete* data. For example, we could count the number of students taking a certain examination or the number of cars in a particular car park. Such figures will be whole numbers (integers).

By measuring, we get data which is *continuous*. In measuring quantities such as temperature, density or current, the accuracy of measurement is limited only by the precision of the measuring instrument.

It is not always necessary to have information which is completely accurate. To illustrate the difficulty of obtaining figures which are precisely correct, consider the membership of a certain trade union. As new members join and others leave, the membership varies. Suppose one of the members dies: it could be several weeks before the news reaches the head office. However, if the union membership was 501 239 last month, it would probably be adequate to say that the current membership is about half a million.

Collecting data by counting may seem easy, but there can be difficulties in practice unless the units to be counted are carefully defined. How many fingers have you on your left hand? Five? Or did you decide that you had four fingers plus a thumb? The difficulty here is that it is possible to count thumbs in a class quite distinct from fingers, but it is equally possible to count fingers and thumbs together as being quite distinct from toes.

To collect information from other people, you must ask the right questions. Suppose you decide to find the average earnings of a group of day-release students. The question 'How much do you earn?' is not sufficiently precise, so we modify it to 'How much do you earn in a week?' so that a period of time is specified. Still the question is not precise enough and is likely to receive responses such as 'Which week?' or 'Should I include overtime payments?' or 'Do you mean take-home pay?'

This illustrates the principle that it is necessary to *define* before you can *classify*. Once you have defined your 'week' and decided what you mean by 'earn', you can start to find out how much each student earns in a week. The collected figures are a set of values of the variable quantity we have defined as 'earnings'. We use the term *variable* for any quantity which can be counted or measured. It should be noted that it is possible to classify certain things on a basis other than measurement where the classification depends on the judgment of an observer. For example, we could separate a group of girls into three

61

classes simply by assessing whether they were blondes, brunettes or red-heads! In this case, the classification is based on an assessment of a particular *attribute*, i.e. hair colour. Similarly, food could be classified as either fit or unfit for human consumption, or milk could be deemed to be fresh or sour.

14.2 Tally diagrams
In a survey of 30 houses, the numbers of rooms in the houses were recorded as follows:

8,	6,	11,	7,	9,	5,	10,	4,	8,	7,
9,	4,	8,	5,	12,	6,	8,	7,	9,	6,
5,	10,	7,	8,	9,	7,	6,	8,	6,	8.

(In counting rooms, all halls, passages, cloakrooms, garages and any places less than 2 m² in area were disregarded.)

How many rooms are there in your home? Check with the figures listed above and find out how many of the 30 houses have fewer rooms than there in your home.

With a set of figures like those above, it is difficult to do any comparison unless the figures are first rearranged in some order. If the numbers are put in order of increasing size and a mark is put alongside for each time any number occurs, the result is a tally diagram as shown below.

No. of rooms	Tally marks	Houses
4	/ /	2
5	/ / /	3
6	ᵗ̶ʰ̶ᵗ̶ /	5
7	ᵗ̶ʰ̶ᵗ̶ /	5
8	ᵗ̶ʰ̶ᵗ̶ / /	7
9	/ / / /	4
10	/ /	2
11	/	1
12	/	1
	Total	30

Note how the fifth tally mark is drawn across the previous four. With large sets of figures, this makes the tally marks easier to total.

The survey was extended to cover 100 houses and this was recorded in the tally diagram shown below, which lists the number of rooms as the variable (x) and the frequency (f) with which each value of the variable occurs.

14.3 Tabulation and grouping of data
Omitting the tally marks from the diagram above leaves a simple table of two columns with values of x and f. Since there are less than a dozen different values for x in this example, grouping is not necessary, but when the range of values for x is very large, the range may be subdivided into smaller groups, usually known as *classes*.

62

No. of rooms (x)	Tally marks	Frequency (f)
4	ЖІ	6
5	Ж ІІІ	8
6	Ж Ж ІІ	12
7	Ж Ж Ж ІІ	17
8	Ж Ж Ж Ж І	21
9	Ж Ж Ж ІІІ	18
10	Ж ІІІІ	9
11	Ж	5
12	ІІІ	3
13	І	0
14	І	1
	Total	100

If 250 students take a certain examination in mathematics, the results may be graded from 0 to 100%. Since this gives so many possible values for x, it is not easy to gauge the performance of the whole 250 and to compare their results with those of last year's students. To simplify the results, the percentage scores can be grouped together into ten grades where each grade corresponds to an interval of 10%. The 250 results can then be tabulated as follows:

Classes	Grade	No. of students
90 to 100%	1	1
80 to 89%	2	6
70 to 79%	3	17
60 to 69%	4	40
50 to 59%	5	62
40 to 49%	6	63
30 to 39%	7	38
20 to 29%	8	16
10 to 19%	9	6
0 to 9%	10	1

When figures are grouped in this way, the class intervals should all be of the same size. Sometimes this is not possible, as in the example above where the class at the top end is slightly larger than the rest. Note that this class contains only one student's mark and therefore the effect of the larger interval for this class is negligible.

Tabulation of continuous data
When the data is continuous, the tabulation is carried out in the same way as with discrete data, but the class intervals should be specified very carefully to ensure that there is neither gap nor overlap between them.

Example   100 roofing tiles are weighed and the weights vary between 0.953 kg and 1.198 kg. Choose appropriate intervals for grouping the results.

The range of weights is from 0.953 to 1.198 kg, i.e. a range of 0.245 kg. It is desirable to divide this into a number of classes, between 5 and 15. The possibilities are 5 classes with interval of 50 g giving 250 g range, 7 x 40 g = 280 g, 9 x 30 g = 270 g, 10 x 25 g = 250 g, or 13 x 20 = 260 g. Of these, the most convenient is likely to be 10 classes with interval 25 g.

These could be listed as 0.950 to 0.975 kg, 0.975 to 1.000 kg, etc., but this would put a value of 0.975 kg into two of the classes, which is impossible, so it is necessary to eliminate overlap at such boundary points. One way of doing this is to specify the classes as follows:

> above 0.950 and up to 0.975 kg
> above 0.975 and up to 1.000 kg

etc. to

> above 1.175 and up to 1.200 kg

Using mathematical symbols, this can be written more concisely as

> $0.950 < x \leqslant 0.975$ kg, etc.

This means that x is greater than 0.950 kg and also x is less than or equal to 0.975 kg (where x is the variable weight).

Exercise E1

1. State in each of the following cases whether the data would be discrete or continuous:
   a) loaves of bread per day sold from a bakery,
   b) daily rainfall figures,
   c) lifetimes of standard heating elements,
   d) number of students per year studying for TEC certificates and diplomas,
   e) crushing strengths of concrete cube test samples.
2. Consider the question 'How many living relatives do you have?' Why is this not a good question for a survey form and how should it be modified in order to get the information required?
3. Comment on the question 'How much sleep do you get each night?'
4. Mark off on a tally diagram the following set of figures for the hours worked by 50 employees:
   40, 39, 42, 40, 40, 39, 43, 37, 40, 41,
   38, 40, 40, 45, 39, 40, 40, 42, 39, 40,
   40, 41, 38, 40, 42, 40, 39, 40, 40, 39,
   42, 40, 40, 44, 40, 38, 40, 47, 40, 41,
   44, 40, 37, 40, 43, 40, 39, 40, 41, 40.
5. From the table in section 14.3 on page 63, find how many students passed the examination by gaining a mark of 40% or more.
6. Explain, in words, the meaning of $15 < x \leqslant 20$.

7. Tally the following set of marks with eight groups of equal interval:

45, 63, 71, 49, 52, 47, 84, 46, 62, 75,
59, 48, 46, 63, 49, 72, 47, 81, 45, 54,
60, 49, 56, 48, 65, 54, 73, 47, 70, 82,
83, 74, 47, 55, 54, 66, 43, 52, 48, 69.

8. Set up a table distributing the following figures between six classes of equal interval. The figures are of resistance measured in ohms.

4.02, 4.97, 5.19, 4.25, 4.76, 4.34, 4.81, 4.70,
5.13, 4.29, 4.04, 5.07, 4.40, 5.15, 4.54, 4.61,
4.17, 4.34, 4.93, 4.72, 4.57, 5.00, 4.88, 4.26,
4.09, 4.93, 4.69, 4.42, 4.77, 4.61. 4.38, 4.60.

How many of these resistances exceed 4.80 ohms?

14.4 Frequency

When we drew the tally diagram for the number of rooms in each of 100 houses, we derived a final column we labelled *frequency* (see page 63). The figures in this column were produced by counting the tally marks. With grouped data the figures in the frequency column give the numbers of items in each class.

A table giving values of a variable quantity (x) and the corresponding frequencies (f) is called a *frequency distribution.*

Frequency diagrams

When a tally diagram would require too many tally marks at the rate of one mark for each item, it may be possible to count the items in tens and put a mark for each ten items. As the number of items increases still further, each tally mark could represent a hundred or a thousand, but this may not be convenient because it does not allow us to show part of a unit.

An alternative method of presentation would be a *pictogram* (*picto*rial dia*gram*), where pictorial symbols are used to represent units. A pictogram to show the number of men employed in a certain industry could look like fig. E1.

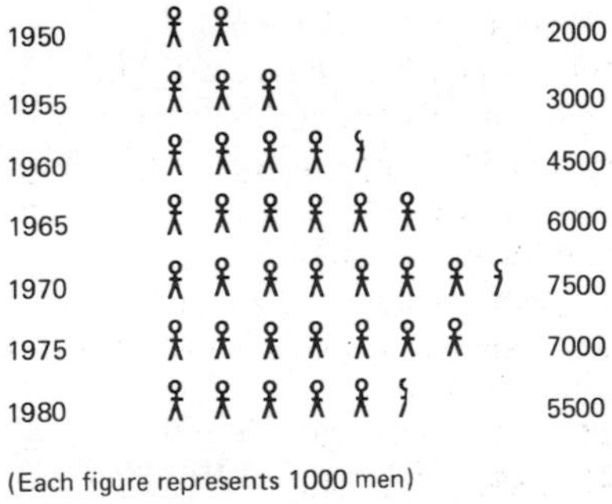

1950	👤 👤	2000
1955	👤 👤 👤	3000
1960	👤 👤 👤 👤 ͻ	4500
1965	👤 👤 👤 👤 👤 👤	6000
1970	👤 👤 👤 👤 👤 👤 👤 ͻ	7500
1975	👤 👤 👤 👤 👤 👤 👤	7000
1980	👤 👤 👤 👤 👤 ͻ	5500

(Each figure represents 1000 men)

Fig. E1 A pictogram

Since it is not always possible to find an appropriate symbol, or for reasons of simplicity, each line of symbols could be replaced by a plain bar of appropriate length to give a *horizontal bar chart* like fig. E2.

65

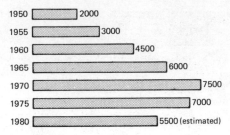

Fig. E2 A horizontal bar chart

Such a bar chart can be drawn either horizontally or vertically and the frequencies can be placed by the bars, as shown above, or marked on a scale along an axis to give a vertical bar chart as shown in fig. E3, but it should be noted that the scale on the vertical axis should start from zero, otherwise the comparison of bar lengths would be misleading since the length of the bar shown would not be proportional to the frequency.

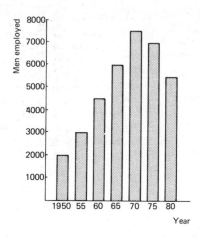

Fig. E3 A vertical bar chart

A bar chart can have two or more sets of bars on the same diagram for purposes of comparison.

Example 1 The bar chart in fig. E4 shows the growth in the numbers of full-time students attending a certain college.

Example 2 A survey of the five main leisure-time activities of 100 young people produced the figures for the bar chart in fig. E5.

66

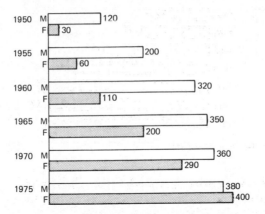

Fig. E4

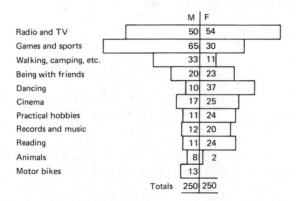

	M	F
Radio and TV	50	54
Games and sports	65	30
Walking, camping, etc.	33	11
Being with friends	20	23
Dancing	10	37
Cinema	17	25
Practical hobbies	11	24
Records and music	12	20
Reading	11	24
Animals	8	2
Motor bikes	13	
Totals	250	250

Fig. E5

Example 3   The table in fig. E6 gives figures (at five-year intervals) for the areas of arable land used for growing grain in the UK. The areas are in thousands of square kilometres.

14.5 Relative frequency

The relative frequency with which any value occurs is given by the frequency of that item divided by the total of the frequencies of all the items.

For a grouped distribution, the relative frequency of any class is given by the frequency of that class divided by the total of the frequencies of all the classes.

Relative frequencies are usually expressed as percentages.

Consider for example the growth of communications in the UK. If the number of letters posted per annum has increased from 10 400 million to 11 000

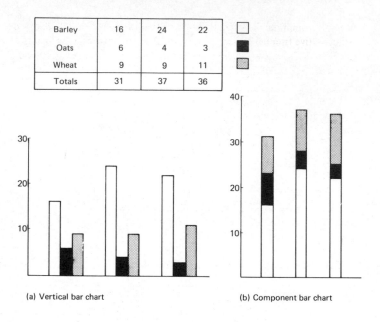

Barley	16	24	22	
Oats	6	4	3	
Wheat	9	9	11	
Totals	31	37	36	

(a) Vertical bar chart   (b) Component bar chart

Fig. E6 UK arable land growing grain (in thousands of km^2)

million in ten years and over the same ten-year period the number of telephone calls has increased from 5600 million to 14 000 million, the change can be seen more clearly in terms of relative frequencies:

	Frequency	Relative frequency	Frequency	Relative frequency
Letters	10 400	65%	11 000	44%
Phone calls	5 600	35%	14 000	56%
Totals	16 000	100%	25 000	100%

From this it can be seen clearly that, although the number of letters posted has increased by 600 million, yet the relative frequency has declined from 65% to 44% because of the greater increase in communication by telephone.

14.6 Relative-frequency diagrams

Some forms of frequency diagram can also be used with relative frequencies by converting the frequencies to percentages and labelling the frequency axis accordingly. This applies particularly to graphs.

Relative frequencies can be illustrated by a *pie chart* in which the sector angles (see section 17.1 and fig. F28) are proportional to the frequencies. Since areas of sectors are proportional to sector angles, it follows that the frequencies are also proportional to the sector areas.

The idea of a component bar chart can be extended to give a *100% bar chart* in which the relative frequencies are proportional to lengths along the bar.

Figures E7 and E8 give a table of values illustrated by both a pie chart and a 100% bar chart.

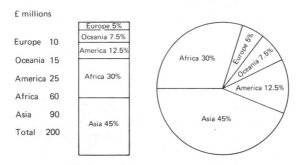

Fig. E7 UK public expenditure on overseas aid

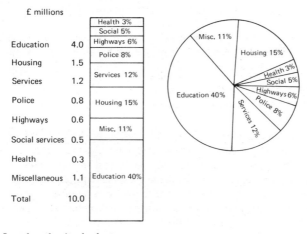

Fig. E8 Local-authority budget

Exercise E2

1. Explain the difference between *frequency* and *relative frequency*.
2. Draw a pictogram to illustrate the following figures for the growth in the number of aircraft in use by a country's airlines:

   1962 300,   1967 400,   1972 450,   1977 500

3. Construct a vertical bar chart to illustrate the following figures giving rainfall in millimetres over a period of six months:

   Jan. 40,   Feb. 32,   Mar. 20,   Apr. 64,   May 88,
   June 45

4. The following figures give monthly average temperatures in degrees Celsius at a certain place. Plot these points on squared paper and join the points by straight lines.

   4.0,   4.2,   6.3,   8.8,   11.9,   15.6,   17.0,   17.6,   15.2,   10.8,   7.2, 5.3

5. Construct a component bar chart to represent the following set of figures, which give areas of land in various parts of the world. The areas are in millions of square kilometres and each unit should be represented by a length of one millimetre.

   Europe 5,   Oceania 8,   S. America 17,   N. America 24,   USSR 20, Asia 26,   Africa 30

6. The percentage of the population of 17-year-olds in different countries in full-time education is as follows:

   Britain 26%,   Denmark 32%,   France 45%,   Japan 75%, Canada 77%,   USA 86%

   Illustrate these figures by a horizontal bar chart.

7. The UK expenditure (in £ millions) on gambling in one year is estimated as 150 on bingo, 200 on football pools, 250 in casinos, 300 in machines and 1100 on horses. Convert these figures to percentages and illustrate by a pie chart.

8. It is estimated in a particular year that the production of commercial vehicles in Britain is divided as follows: light vans $32\frac{1}{2}$%, medium vans $37\frac{1}{2}$%, lorries 25%, with the remainder being of the Land-Rover type. Illustrate these figures by a pie chart.

14.7 Histograms

In some respects a histogram resembles a vertical bar chart, but the two types of diagram should not be confused. A vertical bar chart can represent any quantity which can be measured or counted, and more than one variable may be shown on the same diagram. A histogram shows a frequency distribution of a single continuous variable. A histogram consists of a set of rectangular columns, as does a vertical bar chart, but in a histogram the columns are drawn next to one another without any space between.

Example Having installed new heaters, a horticulturalist checks that the temperature in his glasshouse does not fall below freezing point ($0°C$). The minimum temperatures recorded over 75 nights in mid-winter produced the following distribution, where t is the temperature in degrees Celsius and f is the frequency with which readings fall in each interval.

t	1	2	3	4	5	6	7	8	9	10	11	12
f	0	1	3	7	15	20	13	8	4	2	1	1

The corresponding histogram is shown in fig. E9.

In the above example, all the class intervals were equal and all the columns had the same width, but a histogram can represent a grouped distribution with unequal intervals by columns which have widths proportional to the respective

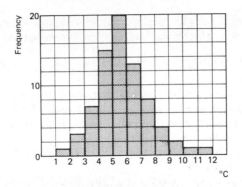

Fig. E9 Histogram showing minimum temperatures

class intervals. If the distribution has unequal intervals, no units should be marked on the vertical frequency axis because frequencies are proportional to the areas of the columns and not to their heights. This implies that the total frequency of the distribution is represented by the total area of the columns.

Example An examination taken by 150 students produced the following distribution of percentage marks, represented by the histogram in fig. E10.

% marks	0 → 25	→ 35	→ 40	→ 45	→ 50	→ 60	→ 70	→ 80	→ 100
Frequency	5	12	16	25	23	29	18	14	8

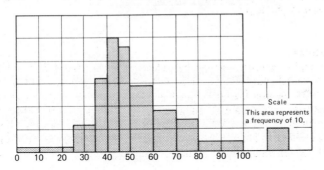

Fig. E10 Histogram showing distribution of marks

For discrete data a vertical bar chart would normally be more appropriate than a histogram, but there are some cases where it is helpful to treat a distribution with a discrete variable as though the variable were continuous. For example, a fish-processing plant for turning herrings into kippers could measure its output either by weighing or by counting the kippers produced. Strictly speaking, weighing would provide continuous data and counting would result in discrete data, but the numbers to be counted would be so large that the production figures could be treated as continuous data whichever method was adopted.

71

When a histogram is drawn to depict a grouped distribution, the class boundaries are marked consecutively along the horizontal axis and, if the distribution has equal class intervals, frequency units are marked on the vertical axis. Changing the frequencies on the vertical scale into percentages changes the diagram to a relative-frequency histogram.

Example   A group of companies analyses its pension scheme for female employees and the following table shows the age distribution of all those who are at present receiving benefit.

Age group	60–64	65–69	70–74	75–79	80–84	85–89	90–95
Frequency %	15	27	24	18	10	5	1

The relative-frequency histogram is shown in fig. E11.

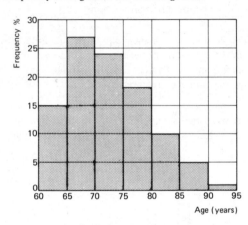

Fig. E11 Relative-frequency histogram showing age distribution of female pensioners

When constructing histograms it is advisable to choose scales such that the height-to-width ratio is between 1:2 and 1:1, since this helps to produce a histogram which looks well-balanced.

Exercise E3
1. In what respects is a histogram different from a vertical bar chart?
2. The following figures give the distribution of times taken by a group of trainees to complete a standard assembly task:

Minutes	8→9	→10	→11	→12	→13	→14
No. of trainees	6	9	14	10	7	4

Illustrate this distribution by a histogram.
3. The percentage moisture content in test samples of concrete after a certain time interval gave the following distribution:

Moisture %	20→22	→24	→26	→28	→30	→32
Frequency	2	6	11	16	12	3

Construct a histogram to illustrate this distribution.

4. The age at entry of a sample of 100 full-time students at a certain college gave a distribution as shown in the table below. (The minimum age for admittance is 16.)

Age in years	up to 16½	up to 17	up to 17½	up to 18	up to 18½	up to 19	over 19
Frequency	8	29	23	21	12	5	2

Draw a histogram to show the distribution of ages in the sample.

5. The following table shows the number of runs scored by a cricketer over 40 innings. Give reasons why a histogram could be unsuitable as a diagram to illustrate such a distribution.

Runs	0	1–5	6–20	21–50	51–100	over 100
Frequency	7	4	5	2	1	1

6. In testing a new variety of tomato plants, the following figures were obtained for the yield in kilograms.

Yield in kg	less than 2.0	<2.5	<3.0	<3.5	<5.0	5 or over
No. of plants	4	11	20	14	5	1

Construct a histogram for this distribution

F   Geometry

15 Angle relationships of parallel and transverse lines

15.1 Measurement of angles

Consider the cartwheel in fig. F1 and the spoke OA in particular. When the cartwheel turns full circle, the spoke OA returns to the horizontal after having

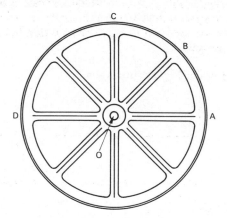

Fig. F1

rotated through one revolution. We say that the angle it has turned through is 360 degrees, i.e. we define a degree as $\frac{1}{360}$ of a revolution. A degree is subdivided into 60 minutes.

∴   1 rev = 360 degrees (360°)

and   1 degree = 60 minutes (60′)

Simplifying our cartwheel diagram to the lines of fig. F2, we can deduce that if OA rotates through 360° in turning through a complete revolution, then it turns through 180° in rotating from OA to OD. Hence angle AOD is 180°.

Similarly, angle AOC is 90° and angle AOB is 45°.

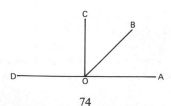

Fig. F2

Note that angle AOB is the angle between the straight lines AO and OB which intersect at O. We say that O is the *vertex* of the angle. Instead of writing angle AOB, we could use the symbol $\angle$ for the word 'angle', or indicate that O is the vertex by drawing a miniature representation of an angle, like an inverted v, over the letter O. Angle AOB can thus be written as $\angle$AOC or AÔC.

$\angle$AOB + $\angle$BOD = 180°, and angles which add up to 180° like this are called *supplementary* angles.

Any triangle must have three straight sides and three angles. In a triangle ABC, the angle ABC is the angle between sides AB and BC, i.e. $\angle$ABC is the angle at vertex B. When there is only one angle at a vertex, it is possible to distinguish the angle simply by using the vertex letter. In this case it is possible to write $\angle$B in place of $\angle$ABC, but this shorter form should be used only when it is quite clear which angle is implied.

Another way of using a single letter to refer to a particular angle is by marking the angle on a diagram with a letter such as x, y or z (or even a Greek letter such as α or θ). We can then refer to angle x and it is clear from the marked diagram which angle is implied.

15.2 Types of angles (fig. F3)

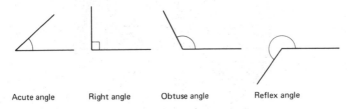

Acute angle   Right angle   Obtuse angle   Reflex angle

Fig. F3 Types of angle

An *acute* angle is between 0° and 90°:

$$0° < x < 90°$$

A *right* angle is exactly 90°:

$$x = 90°$$

An *obtuse* angle is between 90° and 180°:

$$90° < x < 180°$$

An angle of exactly 180° would imply a straight line.

A *reflex* angle is between 180° and 360°:

$$180° < x < 360°$$

15.3 Vertically opposite angles

The angles marked x in fig. F4 are on opposite sides of the same vertex and are equal in size. They are known as a pair of *vertically opposite* angles. The two angles marked y are another pair of vertically opposite angles.

75

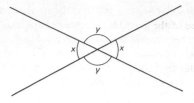

Fig. F4 Vertically opposite angles

15.4 Parallel lines and angle relationships

Figure F5 shows how parallel lines can be constructed by sliding one side of a set-square along a base line or straight edge AB and ruling the lines against another side of the set-square. From this method of construction it is apparent that the angles marked x are equal. Such angles are known as *corresponding* angles.

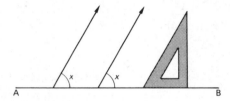

Fig. F5 Constructing parallel lines

In fig. F6 the pair of parallel lines is crossed by another line called a *transversal*. There are four pairs of vertically opposite angles ($a = c, b = d, p = r, q = s$). There are also four pairs of corresponding angles ($a = p, b = q, c = r, d = s$). Since $b = d$ and $d = s$, we deduce that $b = s$. Similarly, $c = p$. Such pairs of angles, on alternate sides of the transversal but between the pair of parallel lines, are called *alternate* angles. The four angles inside the pair of parallel lines (b, c, p, s) are known as *interior* angles and the other four angles (a, d, q, r), being on the outside, are known as *exterior* angles.

Since $b + c = 180°$ and $c = p$, then $b + p = 180°$ also. Similarly, $c + s = 180°$. These pairs of interior angles on the same side of the transversal are therefore supplementary.

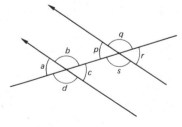

Fig. F6 Parallel lines and transversal

15.5 Identifying parallel lines

Using the converse of the angle relationships in the previous section, we can deduce that two lines must be parallel if a pair of

i)   corresponding angles are equal, or
ii) alternate angles are equal, or
iii) interior angles are supplementary.

Exercise F1

1. In each of the diagrams in fig. F7, find the size of the angle marked x.

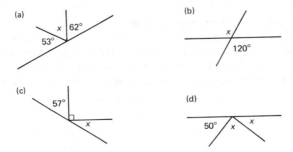

Fig. F7

2. In each of the diagrams in fig. F8, find the values of angles x and y.

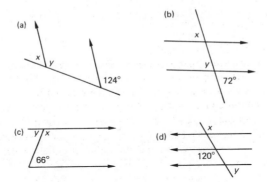

Fig. F8

3. In fig. F9,
    a) which two angles are corresponding angles?
    b) which are alternate angles?
    c) which are both interior and supplementary?

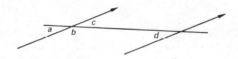

Fig. F9

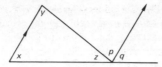

Fig. F10

4. In fig. F10,
 a) which two angles are corresponding angles?
 b) which are alternate angles?
 c) what is the sum of angles x, y and z?

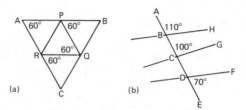

Fig. F11   (a)   (b)

5. In each of the diagrams in fig. F11, pick out which lines are parallel and say why you know they are parallel.

16 Types of triangle and their properties

16.1 The angle sum of a triangle

The sum of the angles of any triangle is $180°$.
 Let the triangle be ABC with angles x, y, z as shown in fig. F12.

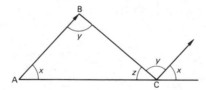

Fig. F12

Draw a line through C parallel to AB.
  The two angles marked x are equal (corresponding).
  The two angles marked y are equal (alternate).
 At C, $x + y + z = 180°$, therefore, in triangle ABC,
  $x + y + z = 180°$
i.e.   the sum of the angles of the triangle is $180°$.
  At C, the angles marked x and y together make an exterior angle. Can you see why this exterior angle is equal to the sum of the angles at A and B? The same principle can be used in every case to show that such an exterior angle is always equal to the sum of the two opposite interior angles.

78

Example In fig. F13, find the values of angles a, b, c and d.

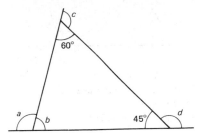

Fig. F13

As an exterior angle, $a = 60° + 45°$

$\therefore$ $a = 105°$

But $a + b = 180°$

$\therefore$ $b = 75°$

As an exterior angle, $c = b + 45° = 75° + 45°$

$\therefore$ $c = 120°$

Also $45° + d = 180°$

$\therefore$ $d = 135°$

Check: the sum of the angles of the triangle is $180°$:

$75° + 60° + 45° = 180°$

What is the sum of the three exterior angles?

$a + c + d = ?$

16.2 Types of triangle (fig. F14)

a) **Acute angled**   All angles are less than $90°$.

b) **Right angled**   One angle is a right angle, i.e. $90°$. The side opposite to this is called the *hypotenuse*.

c) **Obtuse angled**   One angle is greater than $90°$.

d) **Equilateral**   All three sides are equal. All three angles are equal and therefore each is $60°$.

e) **Isosceles**   Two sides are equal. The angles opposite to these sides are therefore equal.

f) **Scalene**   All sides are different in length.

In any triangle, the longest side is opposite the largest angle and, conversely, the shortest side is opposite the smallest angle.

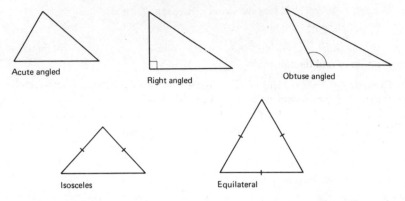

Acute angled

Right angled

Obtuse angled

Isosceles

Equilateral

Fig. F14 Types of triangle

16.3 Complementary angles
Angles which add up to 90° are known as *complementary* angles. In any right-angled triangle, the other two angles must be complementary. Complementary angles have significant connections between their trigonometric ratios (see section 19.6).

The theorem of Pythagoras
This theorem gives us the relationship that the square of the hypotenuse is equal to the sum of the squares of the other two sides. Thus, in fig. F15, for the triangle ABC,

$$a^2 + b^2 = c^2$$

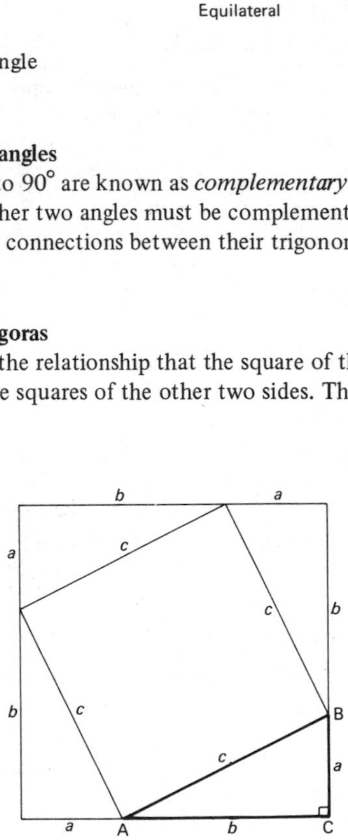

Fig. F15 Pythagoras's theorem

The relationship can be derived from fig. F15 by considering areas:

large square = small square + 4 triangles
$$(a + b)^2 = c^2 + 4(\tfrac{1}{2}ab)$$
$$a^2 + 2ab + b^2 = c^2 + 2ab$$

Subtracting $2ab$ from each side,

$$a^2 + b^2 = c^2$$

If we already know any two sides of a right-angled triangle, we can use this theorem to find the third side.

Example 1   Find the length of the hypotenuse of a right-angled triangle, given that the other two sides are 5 m and 12 m (fig. F16).

$$a^2 + b^2 = c^2$$
$$5^2 + 12^2 = c^2$$
$$25 + 144 = c^2$$
$$c^2 = 169$$
$$c = 13$$

Fig. F16

i.e.   the length of the hypotenuse is 13 m.

Example 2   If the longest side of a right-angled triangle is 3.4 m and the shortest side is 1.6 m, find the length of the third side.

$$a^2 + b^2 = c^2$$

where c is the longest side (the hypotenuse).
Let a be the unknown side, then b is 1.6 m:

$$a^2 + 1.6^2 = 3.4^2$$
$$a^2 + 2.56 = 11.56$$
$$a^2 = 9.00$$
$$a = 3.0$$

i.e.   the length of the third side is 3 m.

16.5 Construction of a right angle
Method 1   From a base AB of length 50 mm, place the point of a compass in turn on either end and draw arcs of respectively 30 mm and 40 mm which intersect at C (fig. F17). Joining C to A and to B gives a right angle at C.

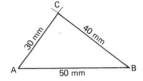

Fig. F17 Construction of a right angle   Fig. F18 Right-angled triangle

It has been known for thousands of years that any triangle with sides in the proportions 3:4:5 will be a right-angled triangle (fig. F18). This fact is often used in practical situations for the construction of a right angle, e.g. when setting out foundations.

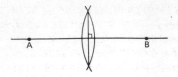

Fig. F19 Construction of a right angle

Method 2   On the base line, select two points A and B equally spaced either side of the place where the line at right angles is required (fig. F19). Placing the point of a compass on each point in turn, and using a radius more than half the distance between A and B, draw arcs of equal radius which intersect both above and below the base line. Joining the points where the arcs intersect gives a line at right angles to AB.

Exercise F2

1. In triangle ABC, $\angle A = 36°$ and $\angle B = 73°$. Find $\angle C$.
2. If $O\hat{P}Q = 41°$ and $Q\hat{O}P = 107°$ in triangle OPQ, find $O\hat{Q}P$.
3. Construct a triangle with sides 60 mm, 80mm and 100 mm and measure the largest angle.
4. Construct a triangle with sides 30 mm, 50 mm and 70 mm and measure the largest angle.
5. For the triangle in fig. F20, prove that the largest angle is a right angle. Which side is the hypotenuse?

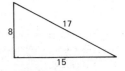

Fig. F20   **Fig. F21**

6. Prove that the triangle in fig. F21 is a right-angled triangle.
   Would a triangle with sides of lengths 14 mm, 48 mm and 50 mm be a right-angled triangle? Give a reason for your answer.
7. Find the hypotenuse of a right-angled triangle if the other two sides are 60 mm and 11 mm in length.
8. Find the length of the third side in fig. F22.

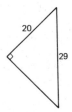

Fig. F22

82

9. A ladder is 5.3 m in length. If one end of the ladder is placed on level ground 2.8 m from a vertical wall, find the vertical height to the point where the other end of the ladder is resting against the wall.
10. A flagpole is held upright by wire stays attached to a point on the pole 5.5 m above ground level. The wire stays are attached to pegs in the ground at a distance of 4.8 m from the base of the pole. Find the length of a wire stay.

16.6 Similar and congruent triangles

For two triangles to be *similar*, they must have the same shape, even though they differ in size. This implies that the angles of one triangle must equal the angles of the other.

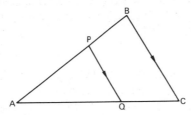

Fig. F23 Similar triangles

In fig. F23,

$\angle APQ = \angle ABC$ (corresponding angles)
$\angle AQP = \angle ACB$ (corresponding angles)

and $\angle A$ is common to both triangles
therefore $\triangle APQ$ is similar to $\triangle ABC$.

When two triangles are similar, the sides of one triangle are proportional to the corresponding sides of the other triangle. Thus, in fig. F23 the ratio of AP to AB is the same as the ratio of AQ to AC and of PQ to BC,

i.e. $\dfrac{AP}{AB} = \dfrac{AQ}{AC} = \dfrac{PQ}{BC}$

For two triangles to be *congruent*, they must be equal in all respects. Not only must the three sides of one triangle be equal in length to the three corresponding sides of the other triangle, and the corresponding angles also be equal, but the two triangles must also be equal in area. When the two triangles are cut out of paper or card, if they are congruent, one will fit over the other exactly.

It is possible to prove that one triangle is congruent to another by confirming that at least three particular parts of one triangle are equal to the corresponding parts of the other, but only certain combinations will satisfy this minimum condition. Three angles equal will not suffice, since two triangles can have corresponding angles equal yet be of different sizes (see fig. F23, in which $\triangle ABC$ and $\triangle APQ$ have three angles equal, but these two triangles are similar, not congruent).

There are four possible combinations by which two triangles can be proved to be congruent by confirming that three parts of one triangle are equal to the corresponding three parts of the other. Using S for side and A for angle, we can write the four acceptable combinations as:

    SSS for three sides equal,
    SAS for two sides and the included angle,
    AAS for two angles and a corresponding side,
    RHS for right angle, hypotenuse and side.

We shall see in section 16.8 that these sets of dimensions are also those which enable us to specify a triangle for construction, i.e. any of these sets of three dimensions will be sufficient for us to draw a triangle of the required size.

Example   If ABC is an isosceles triangle in which AB = BC and D is the mid-point of BC, prove that triangles ABD and CBD are congruent.

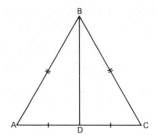

Fig. F24

In fig. F24 we have

    AB = BC   (given)
    AD = DC   (given)

and   side BD is common to both triangles

therefore triangles ABD and CBD are congruent (SSS), which we write as

    $\triangle ABD \equiv \triangle CBD$

The sign $\equiv$ means 'equal in all respects' or 'is identical to'.

Because the two triangles are congruent, the angles opposite to corresponding sides must be equal.

a) Since AD = DC, $\angle ABD = \angle CBD$, i.e. BD bisects $\angle ABC$.

b) Since BD is common, $\angle A = \angle C$, i.e. in any isosceles triangle the base angles are equal.

c) Since AB = BC, $\angle ADB = \angle CDB$. But $\angle ADB + \angle CDB = 180°$, therefore $\angle ADB$ and $\angle CDB$ are both right angles, i.e. BD is perpendicular to AC.

16.7 Finding sides and angles

If two triangles are known to be similar or congruent, the equality of their angles or the relationship between their sides can be used to find an unknown angle or side.

Example   Figure F25 shows a method of estimating the distance between two points on opposite banks of a river. BC is made the same length as CD and $\angle B = \angle D = 90°$.

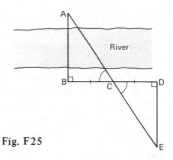

In the triangles ABC and CDE,
  BC = CD   (given)   **Fig. F25**
  $\angle B = \angle D$ (given)
  $\angle ACB = \angle ECD$ (vertically opposite)
∴ the two triangles are congruent (AAS).
  Because the triangles are congruent, we know that AB = ED. If we now measure ED, we can find the distance AB.

The principle of similar triangles can also be used to find the distance between two points either side of an obstruction such as a building or a lake.

Example   Given the dimensions in fig. F26, find the distance between the points D and E.

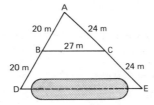

Fig. F26

The ratios AB:AD and AC:AE are both 1:2; thus triangle ABC is similar to triangle ADE and the ratio of BC:DE is also 1:2.
  Since BC is 27 m, DE is 2 × 27 m
i.e.   the distance between points D and E is 54 m.

Further examples of the use of congruency may be found below.

16.8 Constructing triangles
The amount of information which is needed in order to be able to construct a triangle of required shape and size is equivalent to the amount of information needed to prove that one triangle is congruent to another. This means that we can construct the required triangle provided that we know:
a) three sides, or
b) two sides and the angle between them, or
c) two angles and the side between them or a side known to be opposite to one of the given angles, or
d) a right angle, hypotenuse and one other side.

85

Exercise F3

1. ABC is an isosceles triangle in which AB = BC and BD is drawn perpendicular to AC. Prove that triangles ABD and CBD are congruent.
2. If PQR is an isosceles triangle in which PQ = QR, prove that the line which bisects $\angle Q$ also bisects side PR.
3. XYZ is a triangle in which the perpendicular bisector of side XY passes through Z. Prove that XYZ is an isosceles triangle.
4. In the scalene triangle ABC, P is the mid-point of AB, and lines parallel to the other two sides are drawn through P, intersecting AC at Q and BC at R.
   a) Prove that triangles APQ and PBR are congruent.
   b) Prove that triangles APQ and ABC are similar.
5. ABC is a triangle in which AB = 20 m, BC = 15 m, AC = 25 m.
   a) Prove that B = 90°.
   b) Use similar triangles to find the length of the perpendicular from B to AC.
6. If a vertical flag-pole of height 6 m casts a shadow of length 4 m, find the height of a nearby upright telegraph pole which casts a 5 m shadow.
7. In fig. F27, if AB = 12 m and AC = 9 m, find the length of AD.

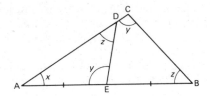

17 Geometry of a circle

17.1 Parts of a circle

With reference to fig. F28, we can identify various parts of a circle and name the curves, lines and areas shown.

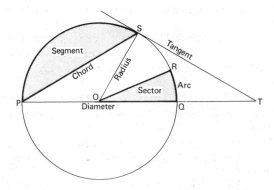

Fig. F28

A *radius* is a straight line drawn from the centre of the circle to its circumference. The plural of radius is *radii*. OP, OQ, OR, OS are all radii.

A *diameter* is a line drawn straight across the circle at its widest part. PQ is a diameter of the circle in the diagram and it should be noted that it passes through the centre of the circle.

(Since it is common practice to talk in general terms of the radius or the diameter of the circle to imply the length of these, it is always advisable to use distinguishing letters when it is necessary to refer to a particular line, e.g. radius OS.)

The *circumference* is the outer edge or perimeter of the circle. The length of the circumference is proportional to the diameter.

An *arc* of the circle is a portion of the curve between two points on the circumference.

A *chord* is a straight line drawn across the circle from one point on the circumference to another. Thus PS is a chord. A chord which passes through the centre of the circle must be a diameter.

In the diagram, line TS is a *tangent* to the circle and it should be noted that it does not cut the circle but merely touches it at S. The distance from T to S is the length of the tangent.

A *sector* of a circle is like a wedge-shaped piece of pie, as can be seen in the diagram where the shaded area OQR is a sector. A sector is bounded on two sides by radii and on the remaining side by an arc. ORS is thus also a sector.

A *segment* of a circle is the portion between a chord and the circumference. The segment shown shaded in the diagram is bounded by the chord PS and the arc between P and S. This segment is the smaller of the two segments into which the circle is divided by chord PS. When such a chord divides a circle into two segments, the larger area is called the *major* segment and the smaller area is the *minor* segment. A diameter divides a circle into two equal segments and then each segment is a *semicircle*.

17.2 Relationship between circumference and diameter

Experiment Take a cylinder (or wheel) of known diameter and start with a mark on its circumference next to a line (A) on the floor. Roll the cylinder along the floor and put a second line (B) where the mark on the cylinder next meets the floor.

Measure the distance AB, which should equal the circumference of the cylinder. By rolling the cylinder on through several revolutions and taking the average, it is possible to find the circumference more accurately. Calculate the ratio of circumference:diameter.

Repeat the experiment with cylinders (or wheels) of several different diameters and determine the circumference:diameter ratio in each case.

This experiment should verify that the ratio of circumference:diameter is a constant. This constant is always denoted by the Greek letter π ('pi'). As a fraction, its closest simple equivalent is $\frac{22}{7}$. As a decimal, for purposes of calculation, we can take

$\pi = 3.14$ or   3.142   or   3.1416   according to the accuracy required.

Since circumference:diameter is equal to π, if we write C for circumference and D for diameter we can derive the simple formula

$C = \pi D$

If r is the radius, so that $D = 2r$, the formula for circumference can be written as

$C = 2\pi r$

Using the circumference formula in one form or the other enables us to solve many problems involving circles.

Example 1   A circular coil of wire of 0.50 m diameter has 20 turns. Find the length of wire in the coil.

$C = \pi D$
$C = 3.14 \times 0.50 = 1.57$ m

If there are 1.57 m in one turn, there are 20 x 1.57 m in 20 turns

i.e.   length of wire = 31.4 m

Example 2   Assuming that kerb-stones of suitable curvature are available in half-metre lengths, find how many would be required to edge a pavement round a 90° bend of 14 m radius.

$C = 2\pi r$
$C = 2 \times \frac{22}{7} \times 14$ m
$C = 88$ m

If a full circle is 88 m, a 90° bend is a quarter circle of 22 m and, if each kerb-stone is $\frac{1}{2}$ m, the number required must be 44.

Example 3   Find the diameter of a tree at the point where its girth is 1.26 m.

$C = \pi D$
1.2 m = 3.14D
$D = \dfrac{1.26 \text{ m}}{3.14} = 0.40$ m

i.e.   the diameter of the tree is approximately 0.4 m.

Exercise F4
1. Find the circumference of a circle with diameter 11 m.
2. Find the circumference of a circle with radius 9 m.
3. Find the diameter of a circle with circumference 18.85 m.
4. If the circumference of the earth is 40 000 km at the equator, find its diameter.
5. If the diameter of a cycle wheel is 650 mm, what is its circumference.
6. A running track is to be constructed with a total length of 400 m. If the two straight stretches are to be 100 m each, find the radius of the semicircular ends.
7. How many turns of wire should there be in a 50 m coil of diameter 400 mm?

17.3 Angles in a circle

In fig. F29, OPQ is an isosceles triangle, since OP = OQ (radii).

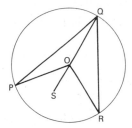

Fig. F29

∴   ∠OPQ = ∠OQP
also   ∠POS = ∠OPQ + ∠OQP   (exterior angle)
∴   ∠POS = 2∠OQP

Similarly, OQR is an isosceles triangle, since OQ = OR (radii).

∴   ∠ORQ = ∠OQR
also   ∠ROS = ∠ORQ + ∠OQR   (exterior angle)
∴   ∠ROS = 2∠OQR

By addition,   ∠POS + ∠ROS = 2∠OQP + 2∠OQR
so   ∠POR = 2∠PQR

i.e.   the angle at the centre is double the angle at the circumference.

Since it is possible to have a diagram in which there may be more than one angle at the centre or circumference, we must define more exactly any such angle by specifying the arc of the circle relating to the angle. With reference to fig. F29, we can define ∠POR as the angle at the centre of the circle subtended by the arc PR. Similarly, we define ∠PQR as an angle at the circumference subtended by arc PR.

We can now rewrite our theorem.

The angle at the centre of a circle, subtended by an arc, is double the angle at the circumference subtended by the same arc.

From this theorem we can derive others which are also important.

Angles at the circumference, subtended by the same arc, are equal.
(See fig. F30. The angles at Q and T are equal since the angle at the centre is twice as large as each of them.)

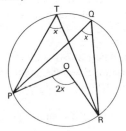

Fig. F30

The angle in a semicircle is a right angle.

(When the arc subtending the angle is a semicircle, the angle at the centre is 180° so the angle at the circumference is 90°.)

17.4 Tangent and radius
It should be noted that the angle between radius and tangent is always a right angle.

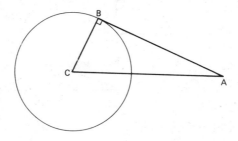

Fig. F31

In fig. F31 if ABC is 90°, then AB must be a tangent to the circle of which C is the centre and CB a radius.

In fig. F32, consider triangles ABC and ADC where AB and AD are both tangents.

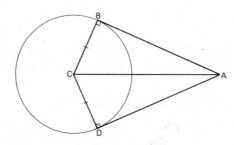

Fig. F32

      B = D = 90°   (angle between tangent and radius)
      AC is common
      CB = CD (radii)
∴   The triangles are congruent (RHS).
      Because the triangles are congruent, AB = AD.
i.e.   *The two tangents to a circle, from any point outside, are equal in length.*

Exercise F5
Find the value of the angle marked *x* in each of the diagrams in fig. F33.

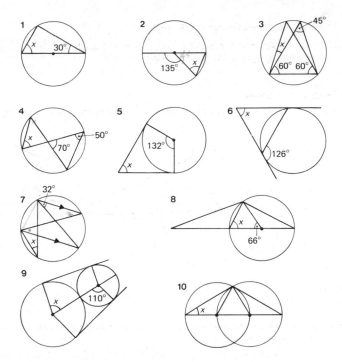

Fig. F33 Exercise F5

17.5 Radians

Angles can be measured either in degrees, minutes and seconds or, alternatively, in radians and milliradians. The relationship between the units is that π radians are equivalent to $180°$,

$$\text{i.e.} \quad 1 \text{ radian} = \frac{180°}{\pi}, \text{ or approximately } 57°18'$$

A complete revolution of $360°$ is 2π radians. Since the circumference of a circle is $2\pi \times$ radius, we can define a radian as the angle subtended at the centre of a circle by an arc equal in length to the radius.

17.6 Degrees and radians

To convert degrees to radians we multiply by π and divide by $180°$.

Example   Convert to radians (a) $45°$, (b) $30°$.
Taking π as 3.1416,

a) $45° \times \dfrac{3.1416}{180°} = \dfrac{3.1416}{4} = 0.7854$ rad

b) $30° \times \dfrac{3.1416}{180°} = \dfrac{3.1416}{6} = 0.5236$ rad

91

Conversion is easier if an electronic calculator is available which will enter an accurate value for π at the press of a button. Alternatively, many books of four-figure tables include a table for the conversion of degrees to radians, or a slide rule could be used.

To convert radians to degrees we multiply by $180°$ and divide by π.

Example   Convert 3.927 radians to degrees.
Taking π as 3.1416,
$$3.927 \text{ radians} = 3.927 \times \frac{180°}{3.1416} = 225°$$

For small angles, measurements are made in milliradians (1 rad = 1000 mrad). Conversions can be made using the following relationships:

$$1° = 17.45 \text{ mrad} \qquad 1 \text{ mrad} = 3'26''$$

17.7 Angular rotation

In section 15.1 we saw that, in one revolution of a cartwheel, an individual spoke rotated through an angle of $360°$. Since $360° = 2\pi$ radians, we see that one revolution is 2π radians. All even multiples of π therefore, in radian measure, correspond to complete revolutions, e.g. 4π radians will be 2 revolutions, 6π radians will be 3 revolutions, etc.

If a shaft or a pulley is rotating at 3 revolutions per second, then the angular rotation must be $3 \times 2\pi$ radians per second. In general terms, a rotation of n revs per second will give an angular velocity of $2\pi n$ radians per second. This is obviously true whether n is a whole number or a fraction.

Example   A shaft is rotating at 50 rev/min. Express this in radians per second.
50 rev/min is 50 revolutions per minute

which is $\dfrac{50}{60}$ revolutions per second

which is $\dfrac{50}{60} \times 2\pi$ radians per second

i.e. $\dfrac{5\pi}{3}$ radians per second

which is 5.236 radians per second.

17.8 Arcs and angles

From the circumference formula $C = 2\pi r$ (section 17.2) we can deduce the arc length corresponding to any angle in radians. Since a circumference length of $2\pi r$ corresponds to one complete revolution of 2π radians, an angle of 1 radian will give a circular arc of length equal to the radius. For a centre angle of θ radians, the arc length at radius r is given by $r\theta$. If the radius is in millimetres, the arc length will also be in millimetres for, in the formula $l = r\theta$, l and r must be in the same units.

Example   A pendulum of length 2.0 m swings through an angle of 0.15 radians in a single swing. Find the length of the arc traced by the pendulum bob.

$l = r\theta$ where $r = 2.0$ m and $\theta = 0.15$ rad
$l = 2 \times 0.15$
$l = 0.30$ m

We can reverse the formula $l = r\theta$ to give $\theta = l/r$, from which we can determine the centre angle corresponding to a given arc length.

Example   Find the angle of lap if 210 mm of a belt drive are in contact with a pulley of radius 150 mm.

$\theta = \dfrac{l}{r}$ where $l = 210$ mm and $r = 150$ mm

$\theta = \dfrac{210}{150} = \dfrac{21}{15} = \dfrac{7}{5} = 1.4$ radians

The principle of angular rotation combined with the formula for the length of an arc enables us to solve problems concerning gear trains and belt drives.

Example   A gear wheel of 200 mm radius is being driven by a second gear wheel of radius 160 mm. Find the angle turned through by the first gear wheel for each revolution of the second.
For the 160 mm gear wheel, $C = 2\pi r = 320\pi$ mm and the 200 mm gear wheel moves through an arc of equal length.

$\theta = \dfrac{l}{r} = \dfrac{320\pi}{200}$

$\theta = 1.6\pi$ radians
$\theta = 288°$

Exercise F6
1. Convert to radians (a) $90°$, (b) $100°$, (c) $72°30'$.
2. Convert to degrees (a) $\frac{2}{3}\pi$ rad, (b) 1.7 rad, (c) 0.435 rad.
3. Convert to mrad (a) $5°$, (b) $2.5°$, (c) $1°45'$.
4. Convert to minutes and seconds (a) 15 mrad, (b) 9.5 mrad, (c) $\frac{1}{7}$ mrad.
5. A pulley is rotating at 1.87 radians per second. Express the speed of rotation in revolutions per minute.
6. The pick-up arm of a record player has an effective length of 230 mm. In playing a record it moves through an arc of 100 mm. Find the angle it moves through, (a) in radians (b) in degrees.
7. Find the angle subtended at the centre of the earth by a section of the equator between two points 10 000 kilometres apart. (Take the length of the equator as 40 000 kilometres.) What should be the time difference between the two points?

18 Areas and volumes

18.1 Quadrilaterals

The name *quadrilateral* means 'four sides'. It is the name given to any closed four-sided plane figure. The diagram in fig. F34 shows a quadrilateral ABCD with a diagonal AC which divides it into two triangles. As the sum of the angles in each triangle is 180°, the sum of the angles of the quadrilateral must be 360°.

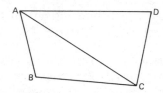

Fig. F34

This can be demonstrated by drawing any quadrilateral on card, then cutting off the four corners and placing them edge to edge on a flat surface so that all four vertices meet at a point. As this can be done without any gap or overlap, the four angles of the quadrilateral are shown to add up to 360°.

The area of the quadrilateral is the sum of the areas of the two triangles.

18.2 Special quadrilaterals

Figure F35 shows some named quadrilaterals with particular properties.

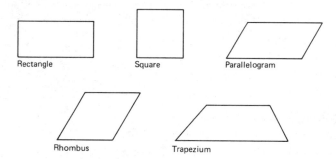

Fig. F35 Special quadrilaterals

Rectangle

All four angles are right angles. Opposite sides are parallel and equal in length. Diagonals are equal in length and bisect one another.

Square

All four angles are right angles. Opposite sides are parallel and all four sides are equal.
Diagonals are equal in length and bisect one another at right angles.

Parallelogram
Opposite angles are equal. Opposite sides are parallel and equal in length.
Diagonals bisect one another.

Rhombus   Opposite angles are equal and are bisected by a diagonal.
Opposite sides are parallel and all four sides are equal.
Diagonals bisect one another at right angles.

Trapezium
One pair of sides is parallel.

18.3 Areas of plane figures
The first section of the table on page 99 gives formulae for the area and perimeter of several common plane figures. The following examples illustrate the
use of these formulae.

a) **Square**   Find the area and perimeter of a square of side 10 mm.
   Area = 10 x 10 = 100 mm^2
   Perimeter = 4 x 10 = 40 mm

b) **Rectangle**   Find the area of a rectangular cross-section 25 mm by 12 mm.
   Area = length x breadth = 25 x 12 = 300 mm^2

c) **Parallelogram**   Find the area of a parallelogram if the two longest sides are
each 25 mm in length and 12 mm apart.

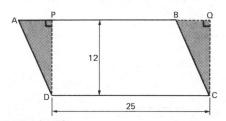

Fig. F36 Area of a parallelogram

The parallelogram is depicted by ABCD in fig. F36. PQCD is a rectangle on
the same base DC. Triangles APD and BQC are congruent (RHS – see section
16.6) and are therefore equal in area. Adding the central quadrilateral to each
of the shaded triangles in turn shows that the parallelogram ABCD is equal in
area to the rectangle PQCD.
i.e.   area = 25 x 12 = 300 mm^2

d) **Circle**   Find the area of a circle of radius 20 mm.
   Area = πr^2 = 3.142 x 20^2 = 1257 mm^2

e) **Semicircle**   Find the area of a semicircle on a diameter of 4 m.
   $D = 2r = 4$ so $r = 2$m
   Area = $\frac{1}{2}\pi r^2 = \frac{1}{2}$ x 3.142 x 2^2 = 6.284 m^2

Applications

Some plane areas can be considered as a combination of two or more standard shapes.

Example 1   The gable end of a house is shown in fig. F37. Calculate the area of brickwork in the gable end.

This figure can be considered as a rectangle of length 8 m and breadth 5 m plus a triangle of base 8 m and height 3 m. Its area is thus

$$lb + \tfrac{1}{2}bh = 8 \times 5 + \tfrac{1}{2} \times 8 \times 3 = 52 \text{ m}^2$$

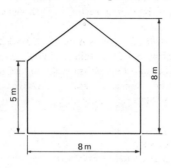

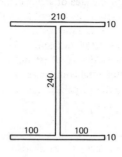

Fig. F37   **Fig. F38**

Example 2   A girder has the dimensions shown in fig. F38. Calculate the area of this cross-section if all units are millimetres.

This cross-section of an I-beam can be regarded as an outer rectangle 260 mm x 210 mm minus two side rectangles each 240 mm x 100 mm. Its area is thus

$$260 \times 210 - 2 \times 240 \times 100 = 6600 \text{ mm}^2$$

Example 3   An archway consists of a rectangular opening topped by a semi-circular arch as shown in fig. F39. Calculate the area of this opening in the wall if the width is 900 mm and the greatest height 2 m.

Fig. F39

The radius of the top semicircle is half the width of 90 mm, therefore
$r = 0.45$ m. The lower rectangular part of the opening is thus 0.90 m wide and
$2 - 0.45$ m high.

Area is $\quad l \times b + \frac{1}{2}\pi r^2 = 0.9 \times 1.55 + \frac{1}{2} \times 3.142 \times 0.45^2 = 1.395 + 0.318$
$$= 1.71 \text{ m}^2$$

Exercise F7

1. Find the area and perimeter of a square of side 42 mm.
2. Find the area of a rectangle 3.4 m x 7.5 m.
3. Calculate the area of a parallelogram if the two longest sides both measure 28 mm and they are 16 mm apart.
4. Construct a square of side 80 mm. Letter the corners A, B, C, D. Mark P as the mid-point of AB and Q as the mid-point of CD. Join A to Q with one straight line and P to C with another straight line..Measure AQ and PC and also the perpendicular distance between them. Use these measurements to calculate the area of the parallelogram APCQ and show that it is half the area of the square.
5. Find the area of a circle of radius 25 mm.
6. If the diameter of a circle is 7 m, find its area.
7. Find the area beneath a semicircular arch of radius 800 mm.
8. A church window fills a space which is a rectangle surmounted by a semi-circle. Find its area if the width is 120 mm and the combined height of rectangle + semicircle is 2.5 m.
9. A square plate of side 36 mm has four holes cut out of it. If the four holes are all circles of diameter 8 mm, calculate the remaining area.
10. If a rectangular slot 32 mm x 6 mm is cut out of the centre of a circle of radius 20 mm, find the area left.

18.4 Volumes

We can find the volume of each of the three-dimensional solids shown in the table on page 99 by using the appropriate formula in each case.

a) *Cylinders*

It is conventional to give the formula for the volume of a cylinder as $\pi r^2 h$, as though the cylinder is standing on a circular end with its long axis vertical. If the cylinder is lying with this axis horizontal, it is equally possible to refer to its length rather than its height h, but this makes no difference to the calculation.

Example   Find the volume of a cylinder with height 2.6 m and diameter 800 mm.
$$V = \pi r^2 h = 3.142 \times 0.40^2 \times 2.6 = 1.3 \text{ m}^3$$

b) *Prisms*

For any prism, the volume is given by multiplying the distance between the parallel end faces by the area of one of them. The end faces may be rectangles, thus giving a rectangular prism, or they may be triangles, which give a triangular prism, but the same principle remains true whatever the shape of the end faces.

Example   Find the volume of a glasshouse which has vertical side walls of height 1 m, a centre height of 3 m, a width of 4 m and a length of 6 m. Each end face is a triangle + rectangle, therefore end-face area is

$$\tfrac{1}{2}bh + lb = \tfrac{1}{2} \times 4 \times 2 + 1 \times 4 = 8 \text{ m}$$

Distance between end faces is 6 m,

∴   volume of glasshouse is $6 \times 8 = 48$ m^3

c) *Cones*

The volume of a cone is one-third of the volume of a cylinder with the same height and base radius.

Example   Find the volume of a cone with base radius 3 m and height 4 m.

$$V = \tfrac{1}{3}\pi r^2 h = \tfrac{1}{3} \times 3.142 \times 3^2 \times 4 = 37.7 \text{ m}^3$$

d) *Pyramids*

As with a cone, the volume is one-third of base area x height. For a square base of side a, this gives $V = \tfrac{1}{3}a^2 h$.

Example   With a square base of side 10 m and height 6 m,

$$V = \tfrac{1}{3} \times 10^2 \times 6 = 200 \text{ m}^3$$

e) *Spheres*

The formula for the volume of a sphere, as shown in the table, is usually written as $\tfrac{4}{3}\pi r^3$. Since $\tfrac{4}{3}$ is a constant, and π is also a constant, we could actually multiply them together and give the formula as $V = 4.19r^3$. One reason why we still use $\tfrac{4}{3}\pi r^3$ is because we can make this as accurate as may be needed by using a sufficiently accurate value for π. A second reason is that it is related to every other formula involving circles, since all of them involve π.

Example   The volume of a ball-bearing of diameter 6 mm (radius 3 mm) is

$$V = \tfrac{4}{3} \times 3.142 \times 3^3 = 113 \text{ mm}^3$$

18.5 Surface areas

The use of the surface-area formulae in the table on page 99 is illustrated by the following examples.

a) *Cylinders*

The curved surface area is given by $2\pi rh$ and the area of each of the two circular ends by πr^2, so the total surface area of a solid cylinder is $2\pi rh + \pi r^2 + \pi r^2$, i.e. $2\pi r(h + r)$.

Example   A quantity of oil drums have to be painted on the outside, including base and lid. Find the surface area of an oil drum which has a radius of 300 mm and a height of 1100 mm.

$$r = 300 \text{ mm} = 0.3 \text{ m} \qquad h = 1100 \text{ mm} = 1.1 \text{ m}$$

∴ $A = 2\pi r(h + r) = 2 \times 3.142 \times 0.3 \times 1.4 = 2.64$ m^2

Title	Figure	Area	Perimeter or circumference
Square		a^2	$4a$
Rectangle		lb	$2(l + b)$
Parallelogram		bh	$2(a + b)$
Triangle		$\frac{1}{2}bh$ $\sqrt{s(s-a)(s-b)(s-c)}$	$a + b + c$ $2s$
Trapezium		$\frac{1}{2}h(a + b)$	$a + b + c + d$
Circle		πr^2 $\frac{\pi D^2}{4}$	$2\pi r$ πD

		Volume	Surface area
Rectangular prism		Area of base × height lbh	Perimeter of base × height $2(lb + bh + hl)$
Cylinder		$\pi r^2 h$ $\frac{1}{4}\pi D^2 h$	$2\pi r(h + r)$ $\pi D(h + \frac{1}{2}D)$
Sphere	Radius r Diameter D	$\frac{4}{3}\pi r^3$ $\frac{1}{6}\pi D^3$	$4\pi r^2$ πD^2
Square pyramid		$\frac{1}{3}$ base × height $\frac{1}{3}a^2 h$	$\frac{1}{2}$ perimeter of base × slant height + base $2al + a^2$
Cone		$\frac{1}{3}\pi r^2 h$ $\frac{\pi D^2 h}{12}$	$\pi r l + \pi r^2$ $\frac{1}{2}\pi D l + \frac{\pi D^2}{4}$

b) *Prisms*

The surface area of a prism consists of the area of the two parallel ends plus the areas of the rectangles connecting corresponding sides.

Example   A glass prism measures 80 mm between parallel 45°, 45°, 90° triangular ends with short sides of 50 mm. Find the total surface area of the five faces of the prism.

There are

two triangles each with area $\frac{1}{2}$ x 50 x 50, total 2 500 mm²
two rectangles each with area 50 x 80.   total 8 000 mm²

The length of the hypotenuse is $\sqrt{(50^2 + 50^2)}$
= 70.7 mm, so the area of the largest rectangle
is 70.7 x 80, i.e   5 656 mm²

∴ the complete surface area of the prism is   16 156 mm²

c) *Hemispheres*

As the surface area of a sphere is $4\pi r^2$, the curved surface area of a hemisphere must be $2\pi r^2$, i.e. double the area of the flat base.

d) *Cones and pyramids*

In each case it is necessary to find the slant height (labelled l in each diagram in the table). The value of l can be found by considering a vertical cross-section through the centre of the cone or pyramid. This gives an isosceles triangle of height h. Taking a symmetrical half of it yields a right-angled triangle and l is the length of its hypotenuse. The value of l is then found by the use of the theorem of Pythagoras.

18.6 Volumes of compound figures

The following examples illustrate how it is possible to find the volumes of some compound shapes by dividing each shape into two or more parts.

Example 1   A rivet has a hemispherical cap of diameter 10 mm and a cylindrical shank 8 mm long and 4 mm diameter. Find the volume of metal in a rivet.

Volume of cap $= \frac{2}{3}\pi r^3 = \frac{2}{3}$ x 3.142 x 5^3 = 261.8 mm³
Volume of shank $= \pi r^2 h = 3.142$ x 2^2 x 8 = 100.5 mm³

Volume of rivet   = 362 mm³

Example 2   A 3 m length of angle iron has an L-shaped cross-section. Regarding this as an outer square 25 mm by 25 mm minus an inner square 21 mm by 21 mm, calculate the volume of metal in this length of angle iron.

Outer square = 25 x 25   = 625 mm²
Inner square = 21 x 21   = 441 mm²

Area of cross-section   = 184 mm²
Volume of bar = 3000 x 184 = 552 000 mm³

100

Example 3   A square nut of side 12 mm and thickness 3 mm has a central hole 6 mm in diameter. Find the volume of metal in the nut (ignoring the screw thread).

Outer square = 12 x 12   = 144 mm²
Inner circle = 3.142 x 3² =   28 mm²

Area of cross-section   = 116 mm²
Volume of nut = 116 x 3 = 348 mm³

As the effect of the screw thread has been ignored, this figure for the answer cannot be accurate to the nearest cubic millimetre, so it would be preferable in this case to give the volume of the nut as 350 mm³ (i.e. correct to two significant figures).

Exercise F8

1. Calculate how many cubic metres of concrete would be required for a 100 m length of strip foundation 400 mm wide and 150 mm deep.
2. Find the volume of a cylinder for which $r = 0.2$ m and $h = 1.5$ m.
3. Calculate the volume of metal in a rod 2.6 m long and 14 mm diameter.
4. What is the volume of a cone with height 3.5 m and radius 3 m?
5. A pyramid has a square base of side 2.4 m and its height is 5 m. Find the volume of the pyramid.
6. Calculate the volume of a sphere 3 m in diameter.
7. If a ball-bearing, 4.00 mm in diameter, is made of metal with a mass of 8.0 mg per mm³, find the mass of the ball-bearing.
8. Find the outer curved surface area of a section of pipe 3 m long and 20 mm radius. Give your answer in square metres.
9. Find the surface area in square millimetres of a ball-bearing of radius 3.00 mm.
10. An insulated wire has a central core of copper wire 2 mm in diameter surrounded by a cylindrical sheath of plastic 2 mm thick. If the mass of copper is 9 mg/mm³ and the mass of the plastic cover is 1 mg/mm³, find the mass of one metre of this insulated wire.
11. In a viscosity experiment, 10 steel balls of 6 mm radius are dropped into a cylindrical glass tube partly filled with liquid. If the internal diameter of the glass tube is 48 mm, find the rise in the level of the liquid when the balls are added.
12. A square nut of side 14 mm and thickness 4 mm has a central hole of 8 mm diameter. If the alloy of which the nut is made has a mass of 8.6 mg/mm³, estimate the mass of the nut.

G   Trigonometry

19 Solving right-angled triangles using sine, cosine and tangent ratios

19.1 Right-angled triangles

For the construction of triangles from minimum data see section 16.8 on page 85.

For the purpose of this section a rough sketch of any required triangle will be adequate, provided that the shape is correct. While it is not necessary to draw such a triangle accurately to scale, care should be taken not to draw two sides equal unless they are known to be so. It is helpful to mark on the sketch of the triangle as much information as possible about the sizes of angles and lengths of sides.

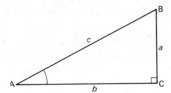

Fig. G1

Figure G1 shows a right-angled triangle ABC. Although the letters A, B and C are put there to label the points where the sides of the triangle intersect (called *vertices*), the same letters may be used to denote the angles. Thus $\angle A$ is the angle at vertex A. It is usual to use the letter a to denote the side opposite to angle A, i.e. side BC. Similarly, side b is opposite to angle B and side c is opposite to angle C.

19.2 Trigonometrical ratios

In a right-angled triangle, the side opposite to the right angle is called the *hypotenuse*, the other sides being called the *opposite* and *adjacent* in accordance with their position relative to one of the two acute angles. In triangle ABC (fig. G1), side BC is opposite to angle A and side AC is adjacent (next to) angle A. Relative to angle B, AC is opposite and BC is adjacent.

Any ratio obtained by dividing the length of one side of a right-angled triangle by the length of a second side will be a trigonometrical ratio. The three most important trigonometrical ratios are defined as follows:

The **sine** of A: $\sin A = \dfrac{\text{opposite}}{\text{hypotenuse}}$

The **cosine** of A: $\qquad \cos A = \dfrac{\text{adjacent}}{\text{hypotenuse}}$

The **tangent** of A: $\qquad \tan A = \dfrac{\text{opposite}}{\text{adjacent}}$

19.3 Trigonometrical tables

Like the four-figure tables in section 6 of squares, reciprocals, logarithms, etc., there are also four-figure tables of trigonometrical ratios. These tables enable us to find the sine, cosine or tangent of any angle in a triangle. Here is a section from a table of tangents.

Natural tangents

Degrees	0′ 0°.0	6′ 0°.1	12′ 0°.2	18′ 0°.3	24′ 0°.4	30′ 0°.5	36′ 0°.6	42′ 0°.7	48′ 0°.8	54′ 0°.9	Mean differences 1′	2′	3′	4′	5′
0	0.0000	0.0017	0.0035	0.0052	0.0070	0.0087	0.0105	0.0122	0.0140	0.0157	3	6	9	12	15
1	0.0175	0.0192	0.0209	0.0227	0.0244	0.0262	0.0279	0.0297	0.0314	0.0332	3	6	9	12	15
2	0.0349	0.0367	0.0384	0.0402	0.0419	0.0437	0.0454	0.0472	0.0489	0.0507	3	6	9	12	15
3	0.0524	0.0542	0.0559	0.0577	0.0594	0.0612	0.0629	0.0647	0.0664	0.0682	3	6	9	12	15
4	0.0699	0.0717	0.0734	0.0752	0.0769	0.0787	0.0805	0.0822	0.0840	0.0857	3	6	9	12	15
5	0.0875	0.0892	0.0910	0.0928	0.0945	0.0963	0.0981	0.0998	0.1016	0.1033	3	6	9	12	15
6	0.1051	0.1069	0.1086	0.1104	0.1122	0.1139	0.1157	0.1175	0.1192	0.1210	3	6	9	12	15
7	0.1228	0.1246	0.1263	0.1281	0.1299	0.1317	0.1334	0.1352	0.1370	0.1388	3	6	9	12	15
8	0.1405	0.1423	0.1441	0.1459	0.1477	0.1495	0.1512	0.1530	0.1548	0.1566	3	6	9	12	15
9	0.1584	0.1602	0.1620	0.1638	0.1655	0.1673	0.1691	0.1709	0.1727	0.1745	3	6	9	12	15
10	0.1763	0.1781	0.1799	0.1817	0.1835	0.1853	0.1871	0.1890	0.1908	0.1926	3	6	9	12	15
11	0.1944	0.1962	0.1980	0.1998	0.2016	0.2035	0.2053	0.2071	0.2089	0.2107	3	6	9	12	15
12	0.2126	0.2144	0.2162	0.2180	0.2199	0.2217	0.2235	0.2254	0.2272	0.2290	3	6	9	12	15
13	0.2309	0.2327	0.2345	0.2364	0.2382	0.2401	0.2419	0.2438	0.2456	0.2475	3	6	9	12	15
14	0.2493	0.2512	0.2530	0.2549	0.2568	0.2586	0.2605	0.2623	0.2642	0.2661	3	6	9	12	16
15	0.2679	0.2698	0.2717	0.2736	0.2754	0.2773	0.2792	0.2811	0.2830	0.2849	3	6	9	13	16
16	0.2867	0.2886	0.2905	0.2924	0.2943	0.2962	0.2981	0.3000	0.3019	0.3038	3	6	9	13	16
17	0.3057	0.3076	0.3096	0.3115	0.3134	0.3153	0.3172	0.3191	0.3211	0.3230	3	6	10	13	16
18	0.3249	0.3269	0.3288	0.3307	0.3327	0.3346	0.3365	0.3385	0.3404	0.3424	3	6	10	13	16
19	0.3443	0.3463	0.3482	0.3502	0.3522	0.3541	0.3561	0.3581	0.3600	0.3620	3	7	10	13	16
20	0.3640	0.3659	0.3679	0.3699	0.3719	0.3739	0.3759	0.3779	0.3799	0.3819	3	7	10	13	17
21	0.3839	0.3859	0.3879	0.3899	0.3919	0.3939	0.3959	0.3979	0.4000	0.4020	3	7	10	13	17
22	0.4040	0.4061	0.4081	0.4101	0.4122	0.4142	0.4163	0.4183	0.4204	0.4224	3	7	10	14	17
23	0.4245	0.4265	0.4286	0.4307	0.4327	0.4348	0.4369	0.4390	0.4411	0.4431	3	7	10	14	17
24	0.4452	0.4473	0.4494	0.4515	0.4536	0.4557	0.4578	0.4599	0.4621	0.4642	4	7	11	14	17

Using this table to find the value of tan 14°33′, we would find from the table

$$\tan 14°30' = 0.2586$$

mean difference for $\underline{\quad 3' \quad} = \underline{\quad 9 \quad}$ (to be added to end figure)

Adding, $\quad \tan 14°33' = 0.2595$

19.4 Finding angles

The same table can be used to find an angle for which the tangent ratio is known.

Example Find the acute angles which have tangents of 0.3000 and 0.1400.
 From the table, tan 16°42′ = 0.3000, so the first angle is easily found. The value 0.1400 does not appear in the table, but the closest value below it is 0.1388, which corresponds to 7°54′.

i.e.   required value = 0.1400
       tan $7°54'$ = 0.1388

       difference =   12

From the mean-difference column along the $7°$ row we see that a difference of 12 is given by an increase of $4'$ in the angle, so the angle must be $7°54' + 4'$ = $7°58'$.

The technique for using tables of sines and cosines is the same, except that in the cosine table (as in the reciprocal table) the figures in the table are decreasing as the angle increases, so the mean differences have to be *subtracted*.

19.5 Some important triangles

Set-squares are usually triangles with angles either $45°$, $45°$ and $90°$, or $30°$, $60°$ and $90°$, i.e. either half a square or half of an equilateral triangle.

In fig. G2(a) we have a square ABCD with a side of unit length and the diagonal AC divides it into two congruent triangles. In triangle ACD, angle A is $45°$ and angle D is $90°$. The diagonal AC of the square is the hypotenuse of the triangle.

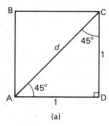

Fig. G2   (a)   (b)

Let the length of AC be d, then, by the theorem of Pythagoras,

$$d^2 = 1^2 + 1^2 = 2$$
∴ $d = \sqrt{2}$

Considering angle A in triangle ACD, we can now see that

$$\sin 45° = \frac{1}{\sqrt{2}} \qquad \cos 45° = \frac{1}{\sqrt{2}} \qquad \tan 45° = 1$$

In fig. G2(b) we have an equilateral triangle of side 2 units with a line QS drawn perpendicular to PR. QS thus divides PQR into two congruent triangles PQS and RQS. Hence S is the mid-point of PR, and PS = SR = 1.

In the right-angled triangle PQS, PQ is the hypotenuse of length 2 units and PS = 1. Putting p for the length of the perpendicular QS, and applying the theorem of Pythagoras, we get

$$p^2 = 2^2 - 1^2 = 3$$
$$p = \sqrt{3}$$

Considering angle P in triangle PQS, we can now see that,

$$\sin 60° = \frac{\sqrt{3}}{2} \qquad \cos 60° = \frac{1}{2} \qquad \tan 60° = \sqrt{3}$$

Considering angle Q in the same triangle,

$$\sin 30° = \frac{1}{2} \qquad \cos 30° = \frac{\sqrt{3}}{2} \qquad \tan 30° = \frac{1}{\sqrt{3}}$$

19.6 Complementary angles

From the trigonometrical ratios we have just considered, we see that

$$\sin 60° = \frac{\sqrt{3}}{2} = \cos 30°$$

and $\quad \sin 30° = \frac{1}{2} \quad = \cos 60°$

The relationships $\sin 60° = \cos 30°$ and $\sin 30° = \cos 60°$ are particular cases of a general rule. This rule involves *complementary* angles, i.e. a pair of acute angles which add up to 90°. For any acute angle θ,

$$\cos \theta = \sin (90° - \theta)$$
and $\quad \sin \theta = \cos (90° - \theta)$

Putting $\theta = 30°$ or $60°$ yields the relationships we have noted above, but θ can have any value between 0 and 90°. The relationships are apparent from a right-angled triangle for, if one of the acute angles is θ, the other acute angle must be $(90° - \theta)$ and the sine ratio of one angle is the cosine ratio of the other (complementary) angle.

19.7 Solution of right-angled triangle

To 'solve' a triangle means to find the size of any unknown angles and the lengths of any unknown sides. There are two basic possibilities.
a) Given two sides, the remaining side may be found by applying the theorem of Pythagoras, and the angles may be found by using sine, cosine or tangent as appropriate.
b) Given one side and one angle, the remaining angle may be found from the fact that the angles are complementary, and the sides may be found by using sine, cosine or tangent as appropriate.

Example 1   In the triangle ABC, right-angled at C, AB is 40 m and BC is 9 m. Find AC and the angles at A and B.

We first draw a small diagram and insert all the given information (fig. G3). From the theorem of Pythagoras,

$$AC^2 = AB^2 - BC^2$$
$$AC^2 = 40^2 - 9^2$$
$$AC^2 = 1519$$
∴ $\qquad AC = 38.97$ m

Fig. G3

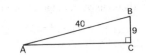

To find angle A, note first that the two given sides are in the positions of hypotenuse and opposite relative to $\angle A$, so we use the sine:

$$\sin A = \frac{\text{opposite}}{\text{hypotenuse}} = \frac{BC}{AB} = \frac{9}{40} = 0.225$$

Using a table of sines we see that

$$\sin 13° = 0.2250$$
$$\therefore \qquad \angle A = 13°$$

To find angle B, note that the two given sides are in the positions of hypotenuse and adjacent relative to $\angle B$, so we use the cosine:

$$\cos B = \frac{\text{adjacent}}{\text{hypotenuse}} = \frac{BC}{AB} = \frac{9}{40} = 0.225$$

From a table of natural cosines we have

$$\cos 77° = 0.225$$
$$\therefore \qquad \angle B = 77°$$

Check: angles A and B should be complementary:

$$13° + 77° = 90°$$

Note that we could have used the calculated length of AC to find either of the angles, but then, had there been an error in calculating AC, it would have affected the result for the angles. It is always better to work for as long as possible directly from the given information.

Example 2   Triangle PQR has an angle of $38\frac{1}{2}°$ at P and a right angle at R. Given that PQ is 16 m, find the lengths of the other two sides.

We first draw a small diagram and insert all the given information (fig. G4).

To find QR, note that QR and the given side PQ are in the positions of opposite and hypotenuse relative to the given angle P, so we use the sine:

$$\sin P = \frac{\text{opposite}}{\text{hypotenuse}} = \frac{QR}{PQ}$$
$$QR = PQ \times \sin P$$
$$= 16 \times 0.6225 \text{ m}$$
$$= 9.96 \text{ m}$$

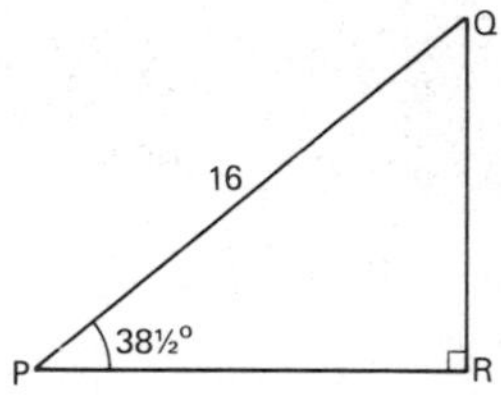

Similarly, to find PR we use the cosine:

$$\cos P = \frac{\text{adjacent}}{\text{hypotenuse}} = \frac{PR}{PQ}$$
$$PR = PQ \times \cos P$$
$$= 16 \times 0.7826 \text{ m}$$
$$= 12.52 \text{ m}$$

Fig. G4

Example 3   ABC is a symmetrical roof truss, the rafters AB and BC each of length 4 m being inclined at an angle of $40°$ to the horizontal. Calculate the span and rise.

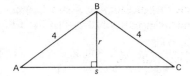

Fig. G5

The roof truss is shown in fig. G5. As it is symmetrical it will be sufficient to calculate the dimensions of one half of it. A perpendicular has been dropped from B on to AC. Denoting the rise by r, and the span by s, we have

$$\sin 40° = \frac{r}{4} \qquad \text{and} \qquad \cos 40° = \frac{\frac{1}{2}s}{4}$$

$$r = 4 \sin 40° \qquad\qquad s = 8 \cos 40°$$
$$r = 4 \times 0.6428 \qquad\quad s = 8 \times 0.7660$$
$$r = 2.57 \text{ m} \qquad\qquad\quad s = 6.13 \text{ m}$$

Deduction of trigonometrical ratios one from another

Method 1   Look up the given ratio in the tables to find the angle, then refer to the appropriate tables to find the new ratio.

Example   Given $\sin x = 0.55$, find $\tan x$.
From sine tables, $\sin 33°22' = 0.5500$ $\therefore$ $x = 33°22'$
From the tangent tables, $\tan x = 0.6586$

Method 2   Draw a small diagram and insert the given ratio. By the theorem of Pythagoras, find the third side of the triangle and hence the required ratio,

Example   Given $\tan x = 1\frac{7}{8}$, find $\cos x$.

$$\tan x = \frac{\text{opposite}}{\text{adjacent}} = \frac{15}{8}$$

These values are put in as shown in fig. G6.
     Now $15^2 + 8^2 = 225 + 64 = 289 = 17^2$
$\therefore$ the hypotenuse is 17 and

$$\cos x = \frac{8}{17}$$

Fig. G6

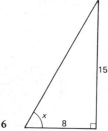

Exercise G1
1. Find the sine, cosine and tangent of each of the following angles: (a) 25°, (b) 62°24', (c) 41°50'.
2. Find the values of (a) sin 28.5°, (b) cos 37.8°, (c) tan 53.75°.
3. Find the acute angle of which the sine is 0.91.
4. Find the acute angle which has a cosine of 0.40.
5. What acute angle has a tangent of 0.6920?

107

6. Given that cos 33°54′ = 0.83, find the angle which has a sine of 0.83.
7. ABC is a triangle right-angled at C. Given that AC is 12 m and BC is 5 m, find AB and angles A and B.
8. Find the smallest angle and the hypotenuse of a right-angled triangle in which the other two sides are 35 m and 12 m.
9. Find the smallest side and smallest angle in a right-angled triangle given that the remaining sides are 40 m and 41 m.
10. A man travels 400 m up a steady slope of 16°. How much higher is he now than when he started?
11. A vertical flagpole casts a shadow of 8 m on level ground when the angle of elevation of the sun is 49°36′. What is the height of the flagpole?
12. The diagonal of a rectangle is 1402 mm and each long side is 1302 mm. Find the length of a short side and the angle between the diagonal and the short side.
13. If ABC is a triangle with a right angle at C, find the values of sin A and cos A by sketching the triangle and using the theorem of Pythagoras, given that tan $A = \frac{3}{4}$.
14. In triangle XYZ, $Z = 90°$ and sin $X = 0.96$. Without using tables, find the values of cos Y and tan Y.

20 Sketching sine and cosine curves

20.1 The sine curve

Figure G7 shows one complete wavelength of a sine curve and it can be seen to consist of two equal and opposite sections, the first half being above the horizontal axis and the second half duplicating it below the axis. This implies that the sines of angles between 0° and 180° are all positive and the sines of angles between 180° and 360° are all negative.

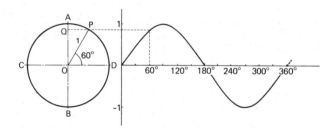

Fig. G7 Construction of a sine curve

The curve is derived from the projection of a rotating unit radius. In the diagram, O is the centre of the circle of which AOB and COD are diameters. The circle is of unit radius, so OP = 1. Think of radius OP as rotating anticlockwise round O as P moves round the circle.

In the diagram, OP is shown at the position where OP makes an angle of 60° with the horizontal. PQ is drawn horizontally to cut the vertical diameter AOB

108

at Q; OQ is thus the *vertical* projection of OP on AOB. The link between the circle diagram and the curve is that every point on the curve is derived from a point on AOB which is given by projecting OP onto AOB when OP makes the appropriate angle with OD. The horizontal dotted line shows how this projection OQ gives a point on the curve when OP has rotated through 60°. The maximum value of the sine curve is +1 and this corresponds to an angle of 90° which is reached when P has travelled round the circle as far as A. When P reaches C, the angle of rotation is 180° and the vertical projection is nil, so sin 180° = 0.

When P arrives at B, the lowest point on the sine curve is reached, corresponding to sin 270° = −1. After completing one revolution, the angle is 360° and sin 360° = 0.

The unit radius OP can continue to rotate about O and this will correspond to further oscillations of the sine curve.

20.2 The cosine curve

A cosine curve can be sketched in a similar way by considering projections of a rotating unit radius onto the *horizontal* diameter.

Comparing the graphs of sin x and cos x in figs G8 and G9, we see that both have the same waveform but there is a difference in displacement of 90° ($\pi/2$ radians).

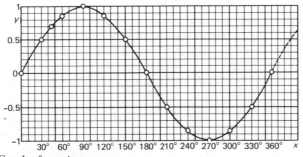

Fig. G8 Graph of $y = \sin x$

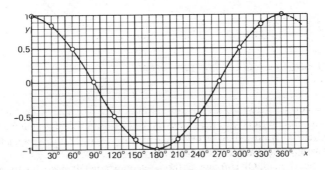

Fig. G9 Graph of $y = \cos x$

109

Revision questions

Revision exercise A

1. Write in figures (a) seventeen thousand five hundred and thirty four, (b) ten thousand and ninety.
2. Write in words (a) 6378, (b) 103 301.
3. What length of cable is left on a 50 metre coil after cutting off three pieces of 4.5 m and four pieces of 2.3 m?
4. Write down the prime numbers between 100 and 110.
5. A secretary comes to work each weekday at 0845 hours and leaves at 1730 hours. If she has an hour each day for lunch and 15 minutes per day for coffee-breaks, find how many hours per week she is at work in the office.
6. A support strut consists of a flat strip of alloy 12.8 mm wide and 100 mm long. The weight is reduced by drilling out a line of holes down the centre. The nine holes are each 8 mm in diameter and they are evenly spaced so that the distance from the edge of any hole to the edge of the strip is the same as the distance from one hole to the next. Find this distance, and also the length of strip at either end past the line of holes. Show all these features on a diagram drawn full-size as for marking out.
7. Substitute the values $x = 7$ and $y = 4$ to demonstrate that $(x + y)^2 - (x - y)^2$ gives the same value as $4xy$.
8. Express 24 minutes as a fraction of an hour in the simplest form.
9. In a concrete mix, the ratio cement : sand : aggregate is 1 : 2 : 4. What fraction of the mix is cement?
10. Components from an assembly line are inspected twice. At the first inspection point one in every nine is found to be faulty and is rejected. At the second quality control point only 95% come up to specification. Find how many components meet specification from an original batch of 2700.
11. An alloy contains 14% of copper. Find the percentage of copper in a new alloy formed by combining 25 kg of the first alloy with 5 kg of pure copper.
12. The ionization constant of boric acid is 0.000 000 000 5s. Express this in standard form.
13. Add together the binary numbers 1011, 1101, and 1001.
14. Simplify $5.2 \times 10^7 + 1.4 \times 10^8$.
15. Evaluate $L(1 + \alpha t)$ when $L = 4500$, $t = 300$, and $\alpha = 0.000\ 024$.

Revision exercise B
Evaluate each of the following by using four-figure tables, giving your answers to three significant figures. Use a calculator to check that your answers are correct.

1. 2.357^2
2. 476.1^2
3. $0.000\ 874^2$
4. $(3.729 \times 10^{-5})^2$
5. $\sqrt{9.247}$
6. $\sqrt{572.6}$
7. $\sqrt{0.006\ 349}$
8. $\sqrt{(8.466 \times 10^{-4})}$
9. $1/7.294$
10. $1/0.020\ 34$
11. $1/(9.273)^2$
12. $\dfrac{1}{4.707} + \dfrac{1}{2.109} + \dfrac{1}{3.206}$
13. $\sqrt{(1.545)^2 + (2.052)^2}$
14. $0.2734 \times 6.392 \times 0.3405$
15. $0.1328 \div 0.0344$
16. $\sqrt{31.24} + \sqrt{46.92} - \sqrt{22.56}$
17. $10^{1.824}$
18. $10^{0.3472} \times 10^{0.6328}$
19. $10^{0.0029}$
20. $10^{-0.512}$

Revision exercise C

1. Simplify $0.25\,(8 - x) + 0.75x - 0.5(x - 4)$.
2. Subtract $3s - 2t$ from $7s + 3t - 5$.
3. Simplify $4.2x^2 + 0.7x^2 - 2.9x^2$.
4. Express as a single fraction $\dfrac{5}{3R} + \dfrac{2}{R} - \dfrac{7}{2R}$.
5. Simplify $(3a^2)^3$.
6. Factorize $ab + ac - bd - cd$.
7. Multiply out $4(3 - x)(2x - 5)$.
8. Solve the equation $3(4t - 1) = 4(t - 3) - 7$.
9. Find the value of x which satisfies the equation
$$1.6(x - 5) = 2(7 + 3.3x) + 3$$
10. Solve the simultaneous equations $3A + 2B = 180°$ and $A - B = 30°$.
11. Find s and t if $s = 75 - 3t^2$ and $3t^2 - s = 21$.
12. If i_1 and i_2 are currents in two sections of an electrical circuit such that $3i_1 + 2i_2 = 13.6$ amperes and $4i_1 - 3i_2 = 0$, find the values of i_1 and i_2.
13. Solve the simultaneous equations
$$\frac{1}{x} + \frac{1}{y} = 7 \quad \text{and} \quad \frac{6}{x} - \frac{1}{y} = 7$$
14. If $V = a + b/I$, find I in terms of a, b, and V.
15. Transpose the formula $V = \dfrac{\pi h^2}{3}(3R - h)$ to make R the subject.
16. Make h the subject of the formula $A = 2\pi r(h + r)$.
17. Transpose $P = mg + \dfrac{mv^2}{r}$ for r.
18. Make α the subject of the formula $\dfrac{L_1}{L_2} = \dfrac{1 + \alpha\theta_1}{1 + \alpha\theta_2}$.

Revision exercise D

1. If a drill with diameter d mm is rotating at n revolutions per second and the product $d \times n$ is a constant 180, draw parallel scales for drills ranging in diameter from 5 mm to 30 mm and corresponding rotational speeds from 36 to 6 revolutions per second.

2. Given that the product of a given wavelength in metres and the equivalent frequency in kilohertz is always 3×10^5, draw parallel scales for long-wave radio broadcasts with frequencies from 350 kHz down to 150 kHz and corresponding wavelengths over the range 850 m to 2000 m. Mark on the scales the appropriate point for BBC Radio 2 broadcast on a wavelength of 1500 m.

3. Write down the equations of the straight lines joining the origin to the points A, B, and C, if A is (6, 3), B is $(-2, -4)$ and C is $(3, -5)$.

4. If P is (4, 3) and Q is $(1, -3)$ find the equation of the line PQ.

5. Using the same axes, draw the straight-line graphs $y = 0.5x + 2$ and $y = 6 - 2x$ for values of x from -1 to $+3$. Find the angle at which the lines cross and the co-ordinates of the point of intersection.

6. The following table gives corresponding values for variables I and V:

I	2	5	8	12	18
V	11.2	13.6	16.0	19.2	24.0

Verify graphically that $V = E + 1R$ and find the numerical values of the constants E and R.

7. Measurements of velocity v after time t are given in the following table:

t	10	12	15	18	20
v	100	88	70	52	40

Plot a graph of v against t and verify that the graph is a straight line. If this corresponds to a relationship $v = u + at$, find the values of the constants u and a. From the graph, estimate the time at which $v = 85$.

Revision exercise E

1. In a radar check, the speed of each passing vehicle is recorded to the nearest kilometre per hour and the following figures are noted for 40 cars:

       40 56 48 67 52 76 78 76 80 69
       45 60 57 69 54 73 81 83 90 86
       47 65 76 73 55 79 86 92 95 84
       53 67 62 68 60 75 79 77 93 66

Tally this set of values with six groups of equal interval, (a) with first group 40–49, (b) with first group 36–45. Compare the results obtained by the two different groupings.

2. An ice-cream manufacturer analyses sales as follows — (a) by number of items: tubs 36%, wrapped bars 24%, ice-lollipops 21%, blocks 19%; (b) by income from sales: tubs 40%, wrapped bars 16%, ice-lollipops 12%, blocks 32%.

    Draw a pie diagram to represent sales by number of items and a second pie diagram to show income. Comment on these two ways of analysing sales.

3. A minibus is in use for 50 weeks in a year and the log shows the weekly mileage covered as follows:

```
122  363  472  532  254  608  325  427  523  245
209  427  316  176  510  275  472  361  167  502
187  249  545  384  491  193  294  554  348  419
405  517  288  444  107  372  571  222  456  605
355  193  348  277  416  509  139  384  266  461
```

Tally this set of values using groups of equal class interval – (a) 6 groups: 101–200, 201–300, etc.; (b) 7 groups: 101–175, 176–250, etc. Find the number of weeks in which the mileage exceeded 450.

4. Analysis of expenditure by a family shows the following sections: housing 30%; food 36%; transport 9%; clothing 12%; entertainment 8%; electricity, gas, telephone, etc. 5%. Illustrate this by constructing an appropriate 100% bar chart.

5. 100 students took a mathematics test (marked out of 50) and the following table shows the distribution of their scores:

Mark	0–9	10–19	20–29	30–39	40–50
Frequency	3	18	42	29	8

Construct a histogram for this distribution.

6. Draw a histogram to illustrate the following distribution.

x	61–70	71–80	81–90	91–100	101–110	111–120
f	12	48	34	20	5	1

Revision exercise F

1. Construct a rectangle 30 mm by 40 mm and label the vertices A, B, C, D. Draw the diagonal AC. Drop perpendiculars BE and DF on to AC.
   (a) Mark on the diagram (with an x) every angle equal to angle CAD.
   (b) Measure the lengths of AC and BE.

2. Given a triangle PQR with three unequal sides and a line SRT through R parallel to PQ, draw a suitable diagram. Label angle QPR as θ and also any other angle of the same size.

3. Find the hypotenuse of a right-angled triangle if the other two sides are 39 mm and 80 mm.

4. If forces of 14 N and 48 N act at right angles to one another, find the magnitude of their resultant.

5. In triangle ABC, $\angle A = 3 \times \angle B = 4x$, $\angle C = 5x$. Find x.

6. Triangles ABC and XYZ are similar. If AB = 4 m, BC = 5 m, CA = 6 m, and XY = 9 m, find YZ and ZX.

7. Tangents AB and AD are drawn to a circle centre C. Prove that AC bisects angle BAD.

8. Given that AD is the perpendicular bisector of BC in triangle ABC, and that the circle on AC as diameter also passes through E such that AD = AE, prove that ABD and ACE are congruent triangles.

113

9. Find the angle of lap if 156 mm of a belt drive are in contact with a pulley of radius 120 mm. Give the answer in both radians and degrees.
10. Six circular holes of diameter 8 mm are drilled out of a rectangular plate 22 mm x 35 mm. Find the area of plate which is left.
11. Calculate the volume of metal in a 4 m length of bar with a semicircular cross-section of radius 12 mm.
12. Find the diameter of a cylindrical drum of height 0.5 m if it is to hold 25 litres (0.025 m³). Find also the area of sheet metal required for its manufacture.
13. Find the volume of metal in a 50 m length of wire of diameter 2 mm.
14. A triangular groove is cut across a bar of flat metal of width 40 mm. If the width of the groove is 3 mm and its depth 2 mm, find the volume of metal removed.

Revision exercise G
1. Find the values of (a) sin 48.6°, (b) cos 39° 28', (c) tan 57.8°.
2. Find the smallest angle and the hypotenuse of a right-angled triangle in which the other two sides are 1.6 m and 6.3 m.
3. From a point 9.8 m outside a circle of radius 5.1 m two tangents are drawn to the circle. Find the length of the tangents and the angle between them.
4. An isosceles triangle has a base of 11.4 m and the other two sides are both 18.5 m. Find the angles of the triangle and the length of the line which divides it into a pair of congruent triangles.
5. Prove that sin 60° + cos 30° has the same value as tan 60°.
6. Triangle OPQ has a right angle at Q. If angle POQ is 46° and OP is 2.30 m, find the length of PQ.
7. Evaluate $IV \cos x$ when I = 13 amperes, V = 240 volts, and $x = \pi/3$ radians.
8. If P is the point (5, 6) and O is the origin, find the angle between the line OP and the x-axis.
9. If a force of 116 N acts at 34° to the horizontal, find its horizontal and vertical components.
10. Triangle ABC is right-angled at C. AB = 3.8 m and AC = 2.9 m. If D is the mid-point of AC, find the length of BD.

Revision exercise (mixed)
1. Find the percentage increase if the population of a village changes from 455 to 546.
2. Find the length of the diagonal and the area of a rectangle with sides 65 mm x 72 mm.
3. Transpose the equation $V = E + IR$ to give I in terms of E, R and V.
4. Express 9/1000 in standard form and 10^{-5} as a decimal.
5. Solve the equation $\frac{1}{2}(2x - 4) - x = 0.4$.
6. Solve the simultaneous equations
   $3x + 4y = 7 \qquad x - 2y = 9$

7. Factorize (a) $3ax + bx - 3ay - by$, (b) $x^2 + 2x + xy + 2y$.

8. Simplify $2\frac{3}{4} + 1\frac{7}{8} - \frac{1}{2}$ of $3\frac{1}{4}$.

9. A formula associated with helical gears is $R = 12c/(Nc - 1)$. Find the value of R when $N = 17$ and $c = 35$ and give your answer as a fraction in its simplest form.

10. If a roof is to slope so that the ratio of rise to span is to be 2:9, calculate the rise of a roof with a span of 3.6 m.

11. Simplify (a) $\dfrac{7^6 \times 7^4}{7^5}$, (b) $\sqrt{\dfrac{4ab^2}{a^3 b^4}}$.

12. Given an angle of $61°\ 37'$, state (a) its supplement, (b) its complement.

13. The formula for the volume of a pipe can be written $V = \pi h (R + r)(R - r)$. Find V when $\pi = 3.142$, $h = 1.5$m, $R = 60$mm, $r = 50$mm.

14. When a certain grade of coal is burnt, it leaves 7% of its mass as ash. If 56 kg of ash per day have to be removed from a boiler house, find the mass of coal being used.

15. Simply $\dfrac{8.4 \times 10^6}{7.0 \times 10^3}$.

16. If paving slabs are made in squares 250 mm × 250 mm, find the number of slabs required to cover an area one metre square.

17. The normal price for a certain wallcovering is £2.40 per roll. For a clearance sale the price is reduced by 15%. Find the sale price.

18. Measurements of mass for five women gave the following values in kilograms:
    57.2   59.6   61.3   63.8   68.6
    Calculate the average.

19. The equation $C = 2\pi r$ (where $\pi = 3.14$) gives the circumference of a circle of radius r. Find the values of C corresponding to values of r in the following table

Radius r in metres	0	2	4	6	8	10
Circumference C in metres	0					

   Plot a graph of C against r over the range of values shown in the table. From the graph, find the radius which corresponds to a circumference of 54 m.

20. In triangle PQR, PQ = QR and S is the mid-point of PR. Prove that triangles PQS and ROS are congruent.
    If ST is drawn perpendicular to PQ, prove that triangles PQS and QST are similar.

21. ABCD is a quadilateral in which angles A, C and D are 80°, 110° and 60° respectively. The bisector of angle B meets AD at E. Calculate angle AEB.

22. Write down the number 19.246 (a) correct to 2 decimal places, (b) correct to 3 significant figures.

23. Simplify $6.2 \times 10^3 - 47 \times 10^2$.

24. How long will it take a tap, dripping at a rate of 800 mm³ per second, to fill a 6 litre can?

25. In tests on a prefabricated panel, the following results were obtained for the deflection (y mm) when loaded with various masses (x kg).

y	8	11	13	17	26	37
x	2	8	12	20	38	60

By a suitable straight-line graph, show that these values are connected by a relationship of the form $y = mx + c$, and find the values of the constants m and c.

26. In fig. RQ1, give appropriate reasons for the following statements:
    a) $\angle a = \angle g$
    b) $\angle n = \angle s$
    c) $\angle m = \angle q = 180°$
    d) $\angle e + \angle j + \angle l = 180°$

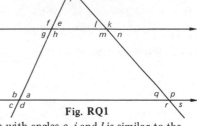

27. In fig. RQ1, list every pair of
    a) alternate angles,
    b) vertically opposite angles,
    c) corresponding angles,
    d) interior supplementary angles.

Fig. RQ1

28. In fig. RQ1, show that the triangle with angles e, j and l is similar to the triangle with angles a, j and q.

29. In fig. RQ2, where OP and OQ are radii of the same circle and R is the midpoint of the chord PQ, prove that triangles OPR and OQR are congruent.

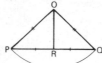

Fig. RQ2

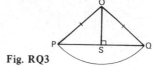

Fig. RQ3

30. In fig. RQ3, OS is drawn perpendicular to chord PQ of a circle centre O. Prove that OS bisects angle POQ.

31. If XYZ is a triangle right-angled at Z, write down the ratios for sin X, cos Y and tan Y in terms of the sides XY, YZ and ZX.

32. Using four-figure tables, write down the values of
    (a) sin 29° 32'   (b) cos 38° 57'   (c) tan 61° 16'.

33. Given that A is an acute angle, find A if sin A = 0.2345.

34. If PQR is a triangle with a right angle at Q, find the values of the angles P and R given that cos P = 0.6789.

35. Explain why the tangent ratio can have values greater than 1.0000, but the sine and cosine ratios can not.

36. ABC is a triangle with a right angle at C. Tan $A = \frac{2}{3}$. Find tan B.

37. Without using tables, slide rule or calculator, find in its simplest form the value of 2 cos 30° tan 60°.

38. If a 30°, 60°, 90° set-square measures 120 mm along its shortest side, find the lengths of the other two sides.

39. A kite-flyer is connected to a fast-moving boat by a line 400 m in length. Find the height of the kite above the water when the line is taut at an angle of 19° to the horizontal.

40. Two tangents are drawn to a circle of 20 mm radius from a point 50 mm from the centre of the circle. Find the length of the tangents and the angle between them.

116

41. In fig. RQ4, prove that angle ACB is a right angle if triangles ACD and CBD are similar.

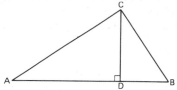

Fig. RQ4

42. The following table shows the number of full-time students in each department of a certain college. Convert these figures to percentages and illustrate them by a pie chart.

    Building   91
    Catering   147
    Commerce   182
    Engineering 161
    Science   119

43. The following figures give the distribution of marks obtained by a group of students in a certain examination:

No. of marks	up to 30	31–40	41–45	46–50	51–60	61–80	81–100
No. of students	9	16	21	24	17	8	5

    Illustrate this distribution by a histogram.

44. (a) Convert 53 to binary form.   (b) Give the equivalent denary number to 101010 on the binary scale.

45. Using four-figure tables, write down the reciprocal, the square and the square root of the number 0.9052.

46. Given that $x = 1.085$, find the values of x^4, $1/\sqrt{x}$, $\log x$.

47. From the formula $y = ah/bx^2$, deduce a formula giving h in terms of the other quantities.

48. Draw a circle 40 mm in diameter and add to it a chord, a tangent and a radius at right angles to the tangent, labelling each part clearly. Calculate the length of the circumference and the area of the circle.

49. In fig. RQ5, prove that the two triangles are similar.

50. In fig. RQ5, prove that the angles marked b and z are equal.

51. Explain the difference between a rhombus and a parallelogram, and prove that the diagonals of a rhombus divide it into four triangles which are all congruent to one another.

52. A gear wheel of diameter 100 mm is driving a second gear wheel of diameter 120 mm. Find the angle the second gear wheel turns through for each revolution of the first. Give your answer in radians and also in degrees.

53. A certain component is a flat strip of alloy 100 mm long and 25 mm wide. Find the percentage decrease in weight if four holes of diameter 16 mm are punched out in line down the centre.

54. In level ground, a drainage trench is excavated 0.4 m wide. The bottom is flat and the sides are vertical, but the depth varies uniformly over a length of 10 m from a depth of 0.6 m at one end to 1.2 m at the other. Find the volume of material dug out of the trench.

55. Find the volume of a cylinder of length 3 m and diameter 0.5 m.

Answers to numerical exercises

Exercise A1 (page 2)
3. 28, 35, 42, 49
4. 54, 63, 72, 81, 90, 99
5. (a) $2 \times 3 \times 7$, (b) 3^5, (c) 2^{10}, (d) $2^4 \times 3^2 \times 11$
6. 23, 29
7. 53

8. 5
9. 48
10. 315
11. 45, 675
12. 28
14. 18 m
15. 9. 8

Exercise A3 (page 11)
1. $\frac{3}{4}$
2. $\frac{3}{5}$
3. $\frac{5}{8}$
4. $\frac{1}{18}$
5. (a) $6\frac{1}{4}$, (b) $1\frac{1}{4}$
6. (a) $7\frac{1}{10}$, (b) $1\frac{7}{10}$
7. (a) 3, (b) $1\frac{1}{3}$
8. 8
9. (a) 4, (b) $1\frac{1}{2}$
10. (a) 875 mm, (b) 640 minutes
11. (a) $3\frac{1}{2}$, (b) $2\frac{1}{2}$
12. $\frac{2}{9}$

13. $\frac{16}{17}, \frac{1}{2}$
14. 320 litres
15. 1200 m
16. $15\frac{1}{4}$ m
17. $4\frac{5}{12}$
18. 4
19. 8
20. $3\frac{1}{4}$
21. $3\frac{1}{2}$
22. $9\frac{1}{3}$
23. $\frac{5}{6}$
24. $\frac{1}{84}$
25. $4\frac{29}{30}$

Exercise A4 (page 16)
1. 0.36
2. (a) $\frac{3}{5}$, (b) $\frac{1}{4}$, (c) $\frac{5}{8}$
3. (a) $\frac{3}{125}$, (b) $\frac{7}{80}$, (c) $\frac{41}{40}$
4. (a) 0.8, (b) 0.8125, (c) 0.44
5. (a) 0.125, (b) 0.343 75, (c) 0.072
6. (a) 0.7, (b) 0.72, (c) 0.571 428
7. (a) 0.81, (b) 0.692307, (c) 0.6428571
8. There are nine altogether.
9. $\frac{5}{9}$
10. $3\frac{1}{7}$
11. (a) 3.14, (b) 0.07, (c) 4.54

12. (a) 64, (b) 0.071, (c) 0.070
13. 6.97
14. 27.5
15. 7.475
16. 4.864
17. 10.32
18. 7.15
19. (a) 12.528, (b) 0.4032
20. (a) 71.332, (b) 2.046
21. 0.712
22. 0.0775
23. 13.43
24. 7.12
25. 2.944

Exercise A5 (page 20)
1. (a) 16, (b) 27, (c) 1000, (d) 3
2. (a) 1, (b) 4, (c) 2, (d) $\frac{1}{4}$
3. (a) 27. (b) 0.5, (c) 0.5, (d) 2.5
4. (a) $2/x$, (b) $1/3x^2$, (c) $2x^3$,
   (d) $2x/3$

5. (a) 5^{3x}, (b) 5^x, (c) $5^x(5^x - 1)$
6. 3
7. 108
8. $2ab^2$
9. 6080
10. (a) $5x$, (b) $9x^2$

Exercise A6 (page 22)
1. 3.75×10^5, 2.173×10^3,
   1.5×10^{-2}, 1.25×10^{-4}
2. 3200, 4 000 000, 0.000 83, 0.0691

3. 7.1×10^5, 7.6×10^{-2}
4. (a) 300, (b) 200
5. 6×10^{38} m

Exercise A7 (page 24)
1. 11, 1001, 11101, 100101, 111011

2. 5, 15, 24, 51, 39
3. 1111, 11011, 110001

Exercise B1 (page 29)
1. 15.80, 22.39, 75.46
2. 1915, 704 100, 47 900 000
3. 0.1781, 0.4670, 0.002 333
4. 1.812×10^7, 78.2×10^{-7}
5. 2.616, 2.301, 2.806
6. 9.419, 22.30, 61.49

7. 0.7000, 0.2214, 0.0070
8. 232.0, 73.36
9. 0.2890, 0.3453, 0.1680
10. 0.338, 0.0024, 0.000 136
11. 1.389, 125, 2110
12. 0.0327, 1.812
13. 0.362, 13.51

Exercise B2 (page 33)
2. 0.6597, 2.6597, $\bar{1}$.5979, 5.9425,
   $\bar{2}$.4775
3. 5.826, 1.229, 17.16, 0.5444,
   0.03987

4. (a) 99.2, (b) 1.83
5. (a) 4.19, (b) 5.61
6. 186
7. (a) 6.12, (b) 16.6
8. 8.26

Exercise B4 (page 39)
1. 11 493
2. 22.606
3. 6542
4. 5.876
5. 133 293
6. 1.7504
7. 6.415
8. 43.48
9. 0.4649

10. 27.42
11. £368
12. $7.50
13. 2.814
14. 3.624
15. 0.907
16. 7.879
17. 15.957
18. 1.30

119

Exercise C1 (page 42)
1.. (a) $4x^2 - x$, $4a + 5b - 7c$,
   (b) $5b$, $3x$, $10x^2$, $12y^3$, 7, $3x^3$,
   $11t^3$, $\frac{1}{2}v^2$
2. a^6, $2b^5$, c^3, $3d^4$, x^{12}, $8y^6$, $\frac{1}{4}$, $1/9z^2$

3. $1 + 4a$, $7b - 10$, $2.3x - 1.2$,
   $y + 2$, $5z^2$
4. $3a + 3b - c$
5. $11/2x$
6. $t^2 + \frac{1}{4}$

Exercise C2 (page 44)
1. $1 + 4x$
2. $(2x + 1)/6$
3. $4x + 2$
4. $3a + 3b$
5. $2a^2 + 2b^2$
6. $3x$

7. $3x(3 - 2x)$
8. $(5 - x)(p - q)$
9. $(a + b)(a + c)$
10. $(2a - 1)(b + 3)$
11. $(x + 2)(7 - y)$
12. $(a + b)(x + y - 1)$

Exercise C3 (page 47)
1. $x = 7, y = 1,\ z = -1$
3. $p = 5, q = 9, r = 2, s = 1, t = 7$
4. $z = 4$

5. $x = 1.6$
6. $a = 4$
7. $x = 3$
8. $x = 2$

Exercise C4 (page 49)
1. $a = 2, b = 1$
2. $x = 5, y = -1$
3. $s = 3, t = 7$
4. $p = -2, q = -3$

5. $x = 5, y = 2$
6. $m = 0.5, n = 1.6$
7. $a = 3, u = 20$
8. $E = 4, R = 3$

Exercise C5 (page 51)
1. $v = 344$
2. $J = 195$
3. $f = \frac{1}{100}$
4. $R = 0.375$
5. $F = 20$

6. $R = 20$
7. $\lambda = V/f$
8. $a = (v - u)/t$
9. $x = (v^2 - u^2)/2a$
10. $V = \frac{1}{6}\pi D^3$

Exercise D2 (page 60)
1. The gradient is 5 and this is the resistance in ohms.
2. The gradient is 0.4.

3. $y = 2x$
4. $K = 0.02$
6. $a = 5, b = 1$
7. $x = 4$

Exercise E1 (page 64)
1. (a) and (d) are discrete, the others continuous.

5. 189
6. x is greater than 15; x is less than or equal to 20.

120

Exercise F1 (page 77)
1. (a) 65°, (b) 120°, (c) 33°, (d) 65°
2. (a) 56°, 124°. (b) 72°, 72°,
   (c) 114°, 66°, (d) 120°, 60°
3. (a) a and d, (b) c and d, (c) b and d
4. (a) x and q, (b) y and p, (c) 180°
5. (a) AB and RQ, BC and PR, AC
   and PQ, (b) BH and DF

Exercise F2 (page 82)
1. 71°
2. 32°
7. 61 mm
8. 21
9. 4.5 m
10. 7.3 m

Exercise F3 (page 86)
5. 12 m
6. 7.5 m
7. 8 m

Exercise F4 (page 88)
1. 34.6 m
2. 56.5 m
3. 6.0 m
4. 12 700 km
5. 2.04 m
6. 31.8 m
7. 40

Exercise F5 (page 90)
1. 60°
2. 67½°
3. 15°
4. 60°
5. 48°
6. 72°
7. 32°
8. 57°
9. 70°
10. 30°

Exercise F6 (page 93)
1. (a) 1.571, (b) 1.745, (c) 1.265
2. (a) 120°, (b) 97° 24′, (c) 24° 55′
3. (a) 87.3, (b) 43.6, (c) 30.5
4. (a) 51′ 34″ or 51′ 35″.
   (b) 32′ 40″, (c) 0′ 29″
5. 17.86
6. (a) 0.4348 rad, (b) 24° 55′
7. 90°, 6 hours

Exercise F7 (page 97)
1. 1764 mm², 168 mm
2. 25.5 m²
3. 448 mm²
5. 1963 mm²
6. 38.5 m²
7. 1 m²
8. 2.85 m²
9. 1095 mm⁵
10. 1065 mm²

Exercise F8 (page 101)

1. $6 \, m^3$
2. $0.19 \, m^3$
3. $4.0 \times 10^{-4} \, m^3$
4. $33 \, m^3$
5. $9.6 \, m^3$
6. $14 \, m^3$
7. $33.5 \, mg$
8. $0.377 \, m^2$
9. $113 \, mm^2$
10. $53.4 \, g$
11. $5 \, mm$
12. $5 \, g$

Exercise G1 (page 107)

1. (a) 0.4226, 0.9063, 0.4663
   (b) 0.8862, 0.4633, 1.9128
   (c) 0.6670, 0.7451, 0.8951
2. (a) 0.4772, (b) 0.7902, (c) 1.3638
3. $65° \, 30'$
4. $66° \, 25'$
5. $34° \, 41'$
6. $56° \, 6'$
7. $13 \, m, \, 22° \, 37', \, 67° \, 23'$
8. $18° \, 56', \, 37 \, m$
9. $9 \, m, \, 12° \, 41'$
10. $110 \, m$
11. $9.4 \, m$
12. $520 \, mm, \, 68° \, 14'$
13. $\sin A = \frac{3}{5}, \, \cos A = \frac{4}{5}$
14. 0.28, 0.28, 0.96

Revision exercise A (page 110)

1. (a) 17 534, (b) 10 090.
2. (a) Six thousand three hundred and seventy eight.
   (b) One hundred and three thousand three hundred and one.
3. $27.3 \, m$
4. 101, 103, 107, 109
5. 37.5 hours
6. $2.4 \, mm, \, 4.4 \, mm$
8. 2/5
9. 1/7
10. 2280
11. $28\frac{1}{3}\%$
12. 5.8×10^{-10}
13. 100 001
14. 1.92×10^8
15. 4532

Revision exercise C (page 111)

1. 4
2. $4s + 5t - 5$
3. $2s^2$
4. $1/6R$
5. $27a^6$
6. $(a - d)(b + c)$
7. $44x - 60 - 8x^2$
8. $t = -2$
9. $x = -5$

10. $A = 48°, B = 18°$
11. $s = 27, t = \pm 4$
12. $i_1 = 2.4$ A, $i_2 = 3.2$ A
13. $x = 0.5, y = 0.2$
14. $I = b/(V - a)$
15. $R = \dfrac{V}{\pi h^2} + \dfrac{h}{3}$
16. $h = \dfrac{A}{2\pi r} - r$
17. $r = \dfrac{mv^2}{P - mg}$
18. $\alpha = \dfrac{L_1 - L_2}{L_2 \theta_1 - L_1 \theta_2}$

Revision exercise D (page 111)
3. $2y = x$, $y = 2x$, $5x + 3y = 0$
4. $y = 2x - 5$
5. $90°; (1.6, 2.8)$
6. 9.6, 0.8
7. $160, -6, 12.5$

Revision exercise E (page 112)
3. 16

Revision exercise F (page 113)
 1. 50 mm, 24 mm
 3. 89 mm
 4. 50 N
 5. $15°$
 6. 11.25 m, 13.5 m
 9. 1.3 rad, $74.5°$
10. 468.4 mm^2
11. 905 cm^3
12. 252 mm, 0.496 m^2
13. 157 cm^3
14. 120 mm^3

Revision exercise G (page 114)
 1. (a) 0.750, (b) 0.722, (c) 1.588
 2. $14.25°$, 6.5 m
 3. 14.0 m, $40°$
 4. $72° 3'$, $72° 3'$, $35° 54'$, 17.6 m
 6. 1.87 m
 7. 1.56 kW
 8. $50.2°$
 9. 96 N, 65 N
10. 2.85 m

123

Revision exercise (mixed) (page 114)

1. 20%
2. 97 mm, 4680 mm^2
3. $I = (V-E)/R$
4. 9×10^{-3}, 0.00001
5. 4.8
6. $x = 5, y = -2$
7. $(x-y)(3a + b), (x + 2)(x + y)$
8. 3
9. $\frac{70}{99}$
10. 0.8 m
11. (a) 7^5 (b) $2/ab$
12. (a) $118°\ 23'$ (b) $28°\ 23'$
13. $0.005\ m^3$
14. 800 kg per day
15. 1.2×10^3
16. 16
17. £2.04
18. 62.1 kg
19. 8.6 m
21. $45°$
22. (a) 19.25 (b) 19.2
23. 1.5×10^3
24. 125 minutes
25. $\frac{1}{2}$, 7
26. (a) alternate
    (b) corresponding
    (c) interior supplementary angles
    (d) angles of a triangle
27. (a) $a = g, b = h, m = p, n = q$
    (b) $a = c, b = d, e = g, f = h,$
    $k = m, l = n, p = r, q = s$
    (c) $a = e, b = f, c = g, d = h, k = p,$
    $l = q, m = r, n = s$
    (d) $a + h = b + g = m + q = n + p$
    $= 180°$
31. YZ/XY, YZ/XY, ZX/YZ
32. 0.4929
36. 1.5
37. 3
38. 208 mm, 240 mm
39. 130 m
40. 45.8 mm, $47°\ 9'$
44. (a) 110101, (b) 42
45. 1.105, 0.8194, 0.9514
46. 1.386, 0.960, 0.0354
47. $h = bx^2y/a$
48. 126 mm, 1257 mm^2
52. 5.236 rad, $300°$
53. 32%
54. $3.6\ m^3$
55. $0.589\ m^3$